D0917071

The Tizard Mission

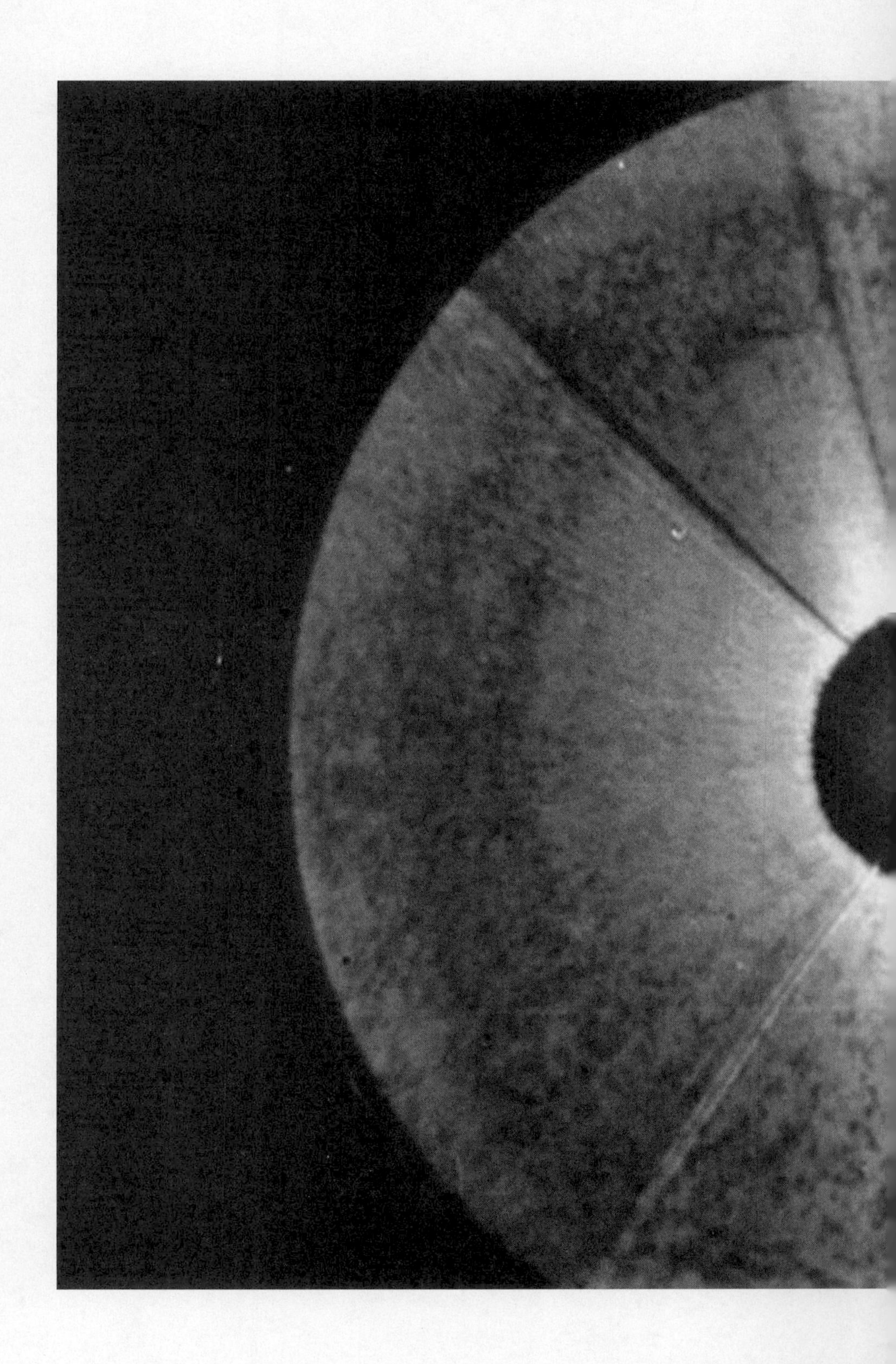

The Tizard Mission

The Top-Secret Operation That Changed the Course of World War II

STEPHEN PHELPS

WESTHOLME
Yardley

Frontispiece: An H2X radar scope image from April 4, 1944, showing a B-17 below. (*482nd Bomber Group/National Archives*)

Westholme Publishing, LLC
904 Edgewood Road
Yardley, Pennsylvania 19067
Visit our Web site at www.westholmepublishing.com

First Printing December 2010
10 9 8 7 6 5 4 3 2 1

ISBN: 978-1-59416-116-2

Printed in the United States of America.

To my father, George Phelps,
who first told me about these things when I was too
young to understand their significance.

It is usually idle to talk of the greatest victory, the greatest general or the greatest invention of a war, these matters are beyond assessment. I suppose, however, that few in a position to judge would hesitate to name the cavity magnetron as having had a more decisive effect on the outcome of the war than any other single scientific device evolved during the war. It was of far more importance than the atomic bomb, which had no effect at all on the outcome of the German War and contributed rather to the shortening of the Japanese War than to its result.

—A. P. Rowe

Contents

Introduction

THIS IS A STORY OF UNLIKELY HEROES: men whose contribution to World War II went beyond the daring exploits of fighter pilots, the stubborn bravery of bomber crews, the tenacity of those who manned the arctic convoys, and even the raw courage of soldiers caught up in hand-to-hand fighting. These people would be hardly recognizable in such company. They worked in the homeland, hundreds, even thousands, of miles from the front line. They wore suits rather than uniforms. Sometimes they even smoked pipes as they went about their work. However, they would make a hazardous sea journey, crossing the U-boat-infested Atlantic with arguably the most precious cargo of the war. What they carried with them would revolutionize the way war was fought. It would save countless thousands of lives—and send many others to their deaths. These men were scientists, Britain's best, and the cargo they carried with them on the liner *Duchess of Richmond* in August 1940, was a small black japanned box. Inside it were the country's top scientific secrets, and the passengers were on their way to America to give them away.

Led by one of the world's great scientific minds, Sir Henry Tizard—aviator, academic and brilliant administrator—their mission was a desperate throw of the dice. They aimed to draw America into the war, or at least to secure the vast manufacturing might of the United States by dazzling the giant neutral nation with a glimpse of the future.

The story of the Tizard mission and the technological revolution it unleashed unfolded during that vital summer and fall of 1940, when Britain stood alone and the world came so close to being shaped into one unrecognizable from that we live in today. It drew on the courage and genius of an extraordinary cast of persons ranging from the president of the United States to a bright young lad from the Welsh valleys, from a bespectacled army officer who'd flown biplanes in World War I to the men who would launch the nuclear age. The Tizard mission brought together all these people, who would play a vital part in creating what was virtually a whole new industry in a matter of months. In the face of opposition and intransigence and under the shadow of Fascist powers striving for world domination through tanks, bombers, and submarines, it was a race against time to invent and manufacture a whole new class of weapons and defensive systems to prevent the destruction of the western democratic way of life.

And all this was made possible when Sir Henry Tizard and his mission slipped into their box of tricks a small metal object no bigger than a man's hand. It would revolutionize not just the war effort but the shape of the world that would emerge from the rubble. A handful of bright men. A lump of metal a few inches across. A secret mission that lasted just a few weeks. In the great sweep of history that is World War II these things might seem insignificant; indeed few are fully aware of the pivotal role it played in the progress of that war. Secrecy, too, contributed to this lack of understanding. The metal object—the cavity magnetron—would eventually render many of the existing weapons systems obsolete. Before I started researching this story the cavity magnetron was unknown to me—but I own one, and so do most people. It powers the microwave oven that is now an integral part of kitchens around the world.

The role of this humble, everyday gadget in the technological war against the Axis powers cannot be overestimated. Picture the history of World War II shaped like an hourglass—huge forces being brought to bear on nations across the world as the story begins to unfold, narrowing sharply as France falls and Britain fights on alone, then mighty outcomes as the free world finally overwhelms the forces of fascism. Now think of the Tizard mission as the neck of that hourglass—a concentration of ideas and invention from the Old World that was released through the New and would power the defeat of the forces of darkness to lay the foundations of the modern world.

ONE

"The Bomber Will Always Get Through"

When Europe descended into war in August 1914, barely a decade had passed since the Wright Brothers' memorable first flight at Kitty Hawk, and just five years since Louis Blériot staggered across the English Channel in his tiny monoplane. Aerial reconnaissance was a new source of information for the generals and a significant contributor to the bloody stalemate on the Western Front. The Great War had seen the emergence of the airplane in the armory of war, and such had been the pace of development that by the early thirties it was clear that the airplane would define any new war. More specifically, the bomber—a fast, efficient machine laden with tons of high explosives and gas that would rain terror from the skies. Men returning from the Western Front after World War I were living warning of the awful effects of the poison gas employed by both sides. Surely that particular terror would now come by airplane to London and the other industrial cities of Great Britain.

Between the wars the power and speed of aircraft continued to develop rapidly, helped by the stimulus of air races such as the Schneider Trophy, in which world air speed records were regularly shattered. And the military were alert to the tremendous capabilities of this new weapon. Bombing had been little more than a sideline for the Royal Flying Corps in World War I, but during the 1920s Britain's newly formed Royal Air Force* was quick to develop aerial bombing as a means of keeping rebel factions in check in the further reaches of the British Empire. Iraq's first taste of aerial bombardment was administered by the British as a means of suppressing tribal forces rising up against the colonial administration. When Hitler, newly installed as chancellor in Germany, began to rearm in defiance of the Treaty of Versailles, the creation of a new and powerful air force was one of his principal aims—and then came the Spanish Civil War. On April 26, 1937, the world woke up to the new and frightening reality of aerial bombing when Germany's Condor Legion, in support of General Franco's attempt to overthrow the newly elected Republican government, attacked the town of Guernica, the ancient and historic capital of the Basques. It was market day, and the town was unusually full, with farmers and peasants making their regular trip into town to buy and sell produce. With no warning several waves of bombers arrived from the south, raining bombs on the defenseless town in an apparently indiscriminate attack. Several hundred people died that day, and three-quarters of the town was destroyed. In just a few hours, as news reports flashed out from northern Spain, the whole world was made painfully aware of the devastation that could be wrought by the modern bomber.

But this devastation was not unexpected in official circles. For some time it had been felt that the bomber had developed into an unstoppable weapon. At least that was the view of many military strategists—and it conditioned the advice they gave to politicians. History is full of men whose influence came to determine the way wars were fought: thinkers whose ideas are seized upon by generals and politicians keen to avoid that trap that awaits them all, fighting *the previous war* instead of this one. In the 1930s one such man was the Italian army general Giulio Douhet, whose theories, published in 1921 under the title *The Command of the Air*, held that all means of

*The RAF was formed by the amalgamation of the Royal Flying Corps and the Royal Naval Air Service on April 1, 1918.

making war other than strategic bombing had become obsolete. Almost the first act of any war, Douhet argued, would be to send vast fleets of bombers to devastate the enemy by raining high explosive from the skies. There might be air battles to secure control of the skies, but once one adversary secured the upper hand the defenders would be powerless to resist the oncoming waves of bombers, which would pound the enemy into submission. Armies would not move, and naval battles would be meaningless in the face of the aerial assault. In Germany these ideas were developed to suggest that civilians, too, were now legitimate targets. In 1935, General Erich Ludendorff published *Der Totale Krieg* (*The Total War*) arguing that no one, not even civilians, could or should be spared by the military.

As the 1930s unfolded Germany, now controlled by Hitler's Nazis, had begun rebuilding its armed forces in clear preparation for war, despite Hitler's protestations of peaceful intentions. Britain, in contrast, seemed desperate to avoid war at any cost. There was not much appetite to embark on another war that threatened to be every bit as bloody as that which had devastated a generation. The pain of this loss had often been exacerbated by the creation of local regiments in which all the young men from, say, a small industrial town in the North would find themselves fighting alongside one another at the front. The policy was designed as a morale booster until that terrible day when they found themselves exposed at the Somme or Vimy Ridge. In one afternoon a community could lose virtually all its young men. The scars that this loss left on the British psyche were enough to ensure a desperate desire to avoid again being drawn into a major European conflict. And there were many in the British ruling classes who seemed to carry a kind of collective guilt about the reparations that were demanded of Germany as it struggled to rebuild its economy. The economic pain increased as the Weimar Republic had come and gone and Germany had fallen victim to massive hyperinflation. Nazi plans to rebuild the economy centered increasingly on rebuilding the military. The development of Britain's armed forces was allowed to lag woefully behind that of Germany. Prime Minister Neville Chamberlain apparently believed that a highly militarized Germany intent on expanding its borders under threat of military sanction would draw the line at taking on the Western powers. Britain would have no need to defend itself against Nazi aggression. Give Germany what it wanted and Britain would never have to face the unstoppable bombers. This was the logic of Chamberlain's policy of appeasement. History would reveal just how misguided it was.

Fortunately, however, there was another current of political thought in Britain in the 1930s, a view of the world that was fundamentally different and was, at the time, confined to the political wilderness. Its principal spokesman was Winston Churchill, who would have to wait several years until Chamberlain's government was finally forced to abandon appeasement, before his star would rise. Chamberlain found himself taking a distinctly ill-equipped country to war, and the illusions Britain still had about its ability to take on the Nazis would soon be exposed.

Defeat could well have been what followed were it not for a small band of scientists and civil servants who had had the foresight to realize that the Nazi threat would not stop at the beaches of northern France, and that if nothing were done, Hitler's bombers would soon be clearing a path for invasion troops on Britain's shores. Among these men were two old friends who first met as students in Germany in the fall of 1908. Frederick Lindemann (later Viscount Cherwell) and Sir Henry Tizard, a physicist and a chemist, respectively, were doing postgraduate work at the laboratory of Walther Nernst at the University of Berlin.

They could not have been more different. Lindemann was rich. Tizard came from a modest background and was anxious about money his whole life. Born in 1885 in the Kent country town of Gillingham, Tizard was as British as they come, middle class with a military background, balding with a wide jaw and bright blue eyes. Lindemann's antecedents were European and American. Born in Alsace, he spoke fluent German, and looked more central European than English. Lindemann played tennis at Wimbledon and was the amateur champion of Sweden. As a teenager he counted Grand Duke

Ernst Ludwig, a grandson of Queen Victoria, among his friends, and he played tennis with the Kaiser and the Tsar.

Tizard was brought up in the English naval town of Chatham and in the suburbs of London, where he attended Westminster, a prestigious English public school. At the age of fourteen he was found to have a blind spot in one eye. It is an irony that though this prevented him following a career in the navy, he would go on to fly airplanes and do much to influence the development of air warfare. The navy it was not to be, but the navy was in his blood. His father had been a naval officer before becoming a naval surveyor and then Assistant Hydrographer to the Royal Navy. He had written the definitive analysis of Nelson's tactics at the battle of Trafalgar and had been made a member of the Royal Society in 1901.

Tizard did well at Westminster School and won a scholarship to Magdalen College, Oxford, where he took a first in chemistry in 1908. For a while it looked as though he might continue his studies at Harvard, but finally it was arranged that he would pursue his particular interest in chemical thermodynamics at the University of Berlin. This would be a defining period of his life, equipping him with a deep understanding of the scientific capabilities of the German nation. In Berlin Tizard saw for the first time the way the Germans were applying themselves to science, and the speed with which it was altering their position in the world.[1]

In the decade or two either side of the turn of the century Germany had become a mecca for science, attracting young scientists like Tizard and Lindemann from across the globe. The first Nobel Prizes were awarded in the first year of the new century. Twenty years later more than half of them had gone to German-speaking scientists: Wilhelm Röntgen for the discovery of X-rays; Adolf von Baeyer for his work on organic dyes; Max Planck for the discovery of energy quanta; and Fritz Haber for the synthesis of ammonia from its elements. Walther Nernst had won his for his work on thermodynamics. There were many others, including the most eminent, Albert Einstein.[2]

Science, and specifically chemistry, had transformed Germany in the nineteenth century from an agrarian society into an industrial and economic powerhouse. A system of technical schools, the

Realschulen, had been established to feed the universities. Above them a further tier of twelve *Technische Hochschulen* had been created specifically for the advancement of science. Soon there was royal patronage with the creation of the prestigious Kaiser Wilhelm Institutes. There were links with industry too, with formal relationships developing between chemists in universities and industrial companies, as well-funded research centers encouraged graduate students to move freely between courses and research programs. The academic environment reflected the cultural traditions of Prussia—hard work, heavy drinking, strict codes of behavior toward women, and even dueling.

It was to this scientific hothouse that Tizard was drawn to study in 1908. The prevailing atmosphere was not simply one of the seemingly limitless possibilities of science; it was also an international melting pot for young scientists from around the world. This spirit of international and interdisciplinary exchange and cooperation would have a profound effect on Tizard as World War II approached.

In Nernst's laboratory, Tizard and Lindemann quickly became good friends, and would remain so for over twenty-five years. But it was a friendship that only went so far, as Tizard himself would later explain:

> There was always something about him which prevented intimacy. He was one of the cleverest men I have ever known. He had been to school in Germany, talked German very well—at least as well as he talked English—and was fluent in French. He was a very good experimenter. He also played games very well. He wanted me to share rooms with him, but I refused. I think my chief reasons for doing so at the time were that he was much better off than I was, and also that we should be speaking English all the time, for he would take no trouble to teach me German.[3]

The size of the gulf in their financial circumstances soon became apparent. Tizard took two rooms in a lodging house, while Lindemann and his brother, a mutual friend, moved into the Adlon, Berlin's finest hotel.

Tizard returned to England in 1909, landing on the day that Blériot made that first crossing of the English Channel by airplane.

He returned to Germany two years later to spend part of the summer with a friend touring the Black Forest, where the inhabitants made such an impression on him that almost half a century later he would record that he had "never been able to think of Germans as wholly bad." He had already formed an underlying affection for Germany that would remain with him throughout the battles he helped wage, against first the Kaiser and then the Nazi regime.

Shortly after his return to England in 1911, Tizard was elected a fellow at Oriel College Oxford at the age of just twenty-six. It was a comfortable life, though one to which he was perhaps ill suited, having a somewhat restless mind. But this pleasant Edwardian idyll did not last long, and with the outbreak of war in 1914 Tizard was soon in uniform, volunteering first for the Royal Garrison Artillery. However, the last few days of peace were spent a long way from the cloistered calm of Oxford and the storm clouds gathering over Europe.

For several years the British Association had had the declared aim of holding one of its annual meetings in Australia, and by the summer of 1914 plans were finally in place. A party of 155 of Britain's leading scientists was ready to sail for Australia, but just three days before the ship was due to leave on July 1, one member dropped out and a last-minute invitation was extended to Tizard. He jumped at the chance and spent the last three months of peace in the company of some of the British Empire's top scientific minds. Here he would make lifelong friendships, and it may have been on this voyage, when he saw what could be achieved by bringing leading scientists together for the exchange of ideas, that the seeds of a future voyage were sown.

In the summer of 1915 Tizard transferred to the Royal Flying Corps as an experimental equipment officer. Here, again, he found himself working alongside Lindemann, who was working on the mathematical theory of recovering an aircraft from an uncontrolled spin. The next few years of war were a fertile period, with both men working on the development of fighting aircraft and aerial tactics. After the war ended both went on to academic posts at Oxford in the 1920s—Lindemann's appointment being on Tizard's recommendation. As the Nazis' aggressive intentions became clear in the following years, Britain's Defense Requirements Committee identified Germany as the "ultimate potential enemy."[4] With their shared background of research in aerial warfare, Tizard and Lindemann both realized how Hitler could use aerial power in pursuit of his goals. And both had the conviction that something could be done about it.

Britain's somewhat defeatist attitude toward an impending war had taken such firm root by the early 1930s that, on November 10, 1932, in a speech to Parliament, Prime Minister Stanley Baldwin issued the ominous warning "the bomber will always get through." It was a perspective that would bedevil Britain's military preparations for the rest of the decade. But it was founded on solid ground—on Britain's geographical position and on improved aircraft and engine design. That Britain is an island has contributed greatly over the centuries to its defense. But the island is small—so small in fact that no point in the country is more than seventy miles from the sea. It had become clear by the early 1930s not only that Germany was becoming a potential aggressor but also that Britain was more vulnerable to bombers than virtually any other country, as a small island with heavy concentrations of both population and industry.

The real question that must be answered in building an air force is what kind of war is envisaged. Is it to be one of aggression—a bomber war—or will it be a defensive war, in which case the fighter will be the principal focus of production and development? Here Britain had a clear orientation.

In the early 1930s Britain's ability to defend against incoming bombers rested on two things. The first was the maintenance of Standing Patrols, small flights of aircraft that were kept aloft with the task of guarding their allotted stretch of the coastline. The second was the information provided by the mainly civilian Observer Corps. Their job was simple, to keep a watchful eye on the skies out to sea and to report accurately the position and bearing of any incoming raiders, information that would then be transmitted to the Standing Patrol. The trouble was, not only were the observers limited by the distance they could see and hear, but as the speeds attainable by the most modern bombers began to climb so the amount of warning that they could give began to fall. The air defense situation looked all but hopeless unless some other solution could be found.

The spur to finding that solution came in 1934 when the RAF staged a mock attack on London to test the capital's air defenses. The results were profoundly shocking. In the first night-time raid the Air Ministry itself was "destroyed," and a subsequent raid managed to "hit" the Houses of Parliament. Air Chief Marshal Sir Arthur Longmore, who was commanding one of the defense zones, wrote that the exercises "had shown that successful interception by fighters of raiding bombers requires more accurate information from the ground as to movements of hostile formations than there was at the time available."[5] Something had to be done.

TWO

America's First Line of Defense

In 1914, as World War I unfolded, President Woodrow Wilson had been determined to keep his country out of what looked increasingly like a ruinous conflict. Wilson described the United States in a speech to Congress as "the one great nation at peace, the one people holding itself ready to play a part of impartial mediation and speak the counsels of peace and accommodation, not as a partisan, but as a friend."[1] Entreaties from Britain and France, who wanted America's weight added to the Allied cause to force an end to the blood-letting, fell on deaf ears. Even when a German U-boat sank the British liner *Lusitania* on May 7, 1915, with the loss of more than a hundred American lives, Wilson remained steadfast in his neutrality. But finally, when Germany began unrestricted submarine attacks on U.S. commercial shipping and tried to form an alliance with Mexico against the United States, Wilson bowed to the inevitable. Now that America was in, this would be a war about the survival of democratic values. In his declaration of war speech on April 2, 1917, Wilson warned that western civilization itself could be destroyed. It would be, he said, a "war to end all wars," and out of it would come an opportunity for a peace that would preclude such wars in the future.

During that conflict, Britain and America had been closely allied and had exchanged information on military technology. Within a few days of the United States entering the war it had dispatched a scientific mission to visit the Allied powers of Great Britain, France, and Italy with a view to obtaining information "on the present status of scientific developments in aid of war." By the end of May a reciprocal Anglo-French mission headed by the eminent physicist Sir Ernest Rutherford was on its way to the United States.

Over the months of June and July 1917, information was exchanged across a wide variety of military technologies, principal among them the pressing problem of submarine detection. German U-boats were posing a huge threat to Allied supply lines, and much work had already been carried out in Britain and France on hydrophones, or underwater listening devices. The French also provided their U.S. counterparts with details of early work on a supersonic transmitting and receiving system that would eventually become sonar. By the following year American scientists had made significant advances on the information they had been given, considerably increasing the effectiveness of British and French countersubmarine measures.

After some local wrangling about control of any continuing exchange, the Americans established a Research Information Service with headquarters in Paris and subsidiary offices in London and Rome. It was to be an offshoot of the Military Committee of the National Research Council, and its job was to send back to the United States a weekly digest of the latest Allied research and development, together with studies of the operational results. But once the war was over this cooperation came swiftly to an end, and in the turbulent decade and a half that followed the armistice the political philosophies of the U.S. and Great Britain started to diverge.

Since the beginning of the century the economic interests of the two nations had been on something of a collision course, as the new economic power of America became a major threat to the dominance of the British Empire. The Great War shifted this movement into a higher gear and the relative positions of the two countries changed dramatically. Britain was forced to liquidate over 10 percent of its overseas assets, becoming a net importer and struggling to attract short-term capital. The United States became a net creditor nation for the first time and the main source of new international investment capital. Inevitably, Britain's balance of trade with the United States tipped sharply in favor of America. As the U.S. grew into a global economic power, President Wilson tried to protect its new strength with a proposal to ensure worldwide peace and prosperity at the end of the war. He wanted a charter to resolve territorial disputes, ensure free trade and international commerce, and establish a League of Nations.

It was a vision for the United States and its position in the postwar world he would prove unable to deliver, when, after the elections of 1918, he encountered a hostile Republican-controlled Senate fearful that membership in the League would restrict the power of Congress to declare war. Led by Henry Cabot Lodge, the Republicans in the Senate balked at one of the central provisions of the League of Nations charter. It required all members to take up arms in defense of any other signatory against aggression of any kind. Lodge refused to sign on to what he saw as an open-ended commitment to send soldiers into conflict regardless of the national interests of the United States. But a determined Wilson was unwilling to compromise; when his stubbornness failed to overcome the resistance of the Senate, the U.S. rejected the Treaty of Versailles, and the League of Nations came into being without the nation that had done so much to promote it. Hamstrung from the start, it found that its mission to oversee the peace was fatally undermined. America had turned inward, more interested in itself than in Wilson's new world order.

In 1919, Henry Cabot Lodge articulated his vision for America's future, a vision that would lay the foundation for the isolationism of the next two decades:

> The United States is the world's best hope, but if you fetter her in the interests and quarrels of other nations, if you tangle her in the intrigues of Europe, you will destroy her powerful good, and endanger her very existence. Leave her to march freely through the centuries to come, as in the years that have gone. Strong, generous, and confident, she has

> nobly served mankind. Beware how you trifle with your marvellous inheritance; this great land of ordered liberty. For if we stumble and fall, freedom and civilization everywhere will go down in ruin.[2]

It was an affirmation of clear and simple American values which were not to be put at risk by involvement in the dark machinations of the Old World. America had washed its hands of Europe. Germany had been put back in its place, and the old balance of power in Europe restored; under the League of Nations the European colonial powers would keep each other in check. America's growing economic power would surely be enough to deter any potential aggressor. As the troubled peace on that continent began to unravel, the United States settled into isolation, reassured by the vast expanse of ocean that separated it from the squabbles and strains of the rest of the world. From 1919 onward America determined never again to be drawn into someone else's war.

At the same time, America was steadily losing its nostalgic ties to what had been its mother country. A huge influx of immigrants had changed the ethnic makeup of the country, and even the heartland of the Midwest was increasingly populated with farmers of Scandinavian, German, or Irish descent who felt no affiliation to Britain. With the stock market crash of 1929 and the Great Depression that followed, domestic concerns far outstripped foreign policy considerations in the minds of both politicians and public. The imposition of high tariffs on imports under the Smoot-Hawley Tariff Act, passed in June 1930, further separated the United States and its citizens from outside influences, and other countries responded in kind. In 1931, Britain instituted a new system of trade tariffs apparently designed to protect trade with the Empire and Dominions. It pulled out of the international gold standard in October 1931 to create a controlled "Sterling Area." In fact these were just desperate measures with which the British hoped to tackle the domestic effects of the Depression, a slump made worse by protectionist tariffs. To many in the United States, however, they seemed designed primarily to block American exports. As Franklin D. Roosevelt succeeded Herbert Hoover, public opinion pressed the new president to address the nation's economic difficulties above all else.

As autocracies began their rise in Europe and the Far East, in the United States the capitalist system itself was viewed with increasing skepticism. In the era of prohibition and gangsters, wealthy bankers were referred to as "banksters." But the sharpest vitriol was reserved for the arms trade. With the arrival of the 1930s, arms manufacturers were increasingly regarded as having promoted the Great War in the interests of their own profit, and a series of books and articles appeared with inflammatory titles like *Iron, Blood, and Profits* and *Merchants of Death*, which railed against the avarice of the munitions firms, alleging it to be the principal cause of war. After a number of scandals broke revealing that arms manufacturers had disseminated false information about the growing strength of rearmament around the world, in 1934 a congressional committee led by Senator Gerald Nye began a series of hearings to investigate the munitions business. For two years of widely publicized deliberations, the Nye Committee identified "an efficient, callous, international combine" that had "compromised the integrity of public servants, undermined disarmament, ignored embargoes, lobbied for ever-increasing armament expenditures, plotted war scares, and in wartime dealt with all belligerents regardless of moral or legal position."[3] It also identified British machinations which it claimed had drawn a naïve and idealistic America into the war against its own interest. The committee documented the huge profits that arms factories had made during the war. The British, it found, had benefited enormously from American largesse. Between 1915 and January 1917, the United States had loaned the United Kingdom and its allies some $2.3 billion. Not only had more than one hundred thousand Americans died in someone else's war, they had paid for the privilege.

This strong antiwar sentiment coalesced in a series of Neutrality Acts passed by Congress during the 1930s. U.S. policy in World War I had been to aggressively assert its right to trade with any belligerent country. But with hindsight this had been recognized as a policy that made a slide into war inevitable. This time a different approach was called for, and between August 1935 and May 1937, Congress passed three separate Neutrality Acts. Designed to insulate the United States from future wars, they imposed an embargo first on the sale of munitions to all belligerents, then a ban on the provision of loans, and finally a prohibition against U.S. citizens traveling on belligerents' passenger vessels and against American vessels carrying arms to belligerents. Though something of a blunt instrument (the term "belligerent" made no distinction between aggressor and victim), the acts were a popular reflection of prevailing public opinion, particularly among Republicans and conservative southern Democrats. Support for the acts was never so great in Congress that Roosevelt could not have used his presidential veto, but needing Congressional support on domestic issues, and facing reelection in 1936, he decided not to risk alienating these powerful forces in Congress and the country at large, and reluctantly signed them into law.

In Britain things were seen rather differently. As the Nazi threat loomed in Europe in the 1930s and isolationism gripped American popular opinion, the British saw the Nazis' ideology as endangering democracies at large, and so were looking to the United States for support on the basis of American self-interest if nothing else. Roosevelt had no love for Hitler's regime. But throughout the 1930s he made a clear distinction between the German people and the Nazis. Hitler and his National Socialists only came to power, he felt, because the country was in such a chaotic state economically and socially after Versailles and the failure of the Weimar Republic. He later recalled, "when this man Hitler came into control of the German government Germany was busted, . . . a complete and utter failure, a nation that owed everybody, disorganized, not worth considering as a force in the world."[4] National Socialism might be a short-term fix, but he felt certain it was economically doomed in the long run. The Nazis' military ambitions would sooner or later bankrupt them.

Nonetheless the harsh fact was that in Germany in the mid-1930s rearmament seemed to be working its economic magic. Roosevelt noted in 1936, "there is no-one unemployed in Germany, they're all working in war orders," though he went on to say "eventually, of course, they will have to pay for it."[5] He was aware that, like a gambler who constantly doubles down in order to retrieve his losses, the Nazi economy could only be maintained by constant expansion. The logic of it all was that the Nazis were aiming for world domination—

and the great powers of the western hemisphere were by no means immune.

This was the mindset with which Roosevelt approached the developing situation in Europe. A naval man by training, the president subscribed to the influential doctrines of Admiral Alfred Mahan, the great military strategist. Sea power, and the vast expanses of the oceans that separated the United States from the belligerent worlds of Europe and Asia, were the keys to the country's security. But Mahan was clear on one crucial point—that the U.S. Navy could not defend both the Atlantic and the Pacific at the same time. In the mid-1930s, as Japan's aggressive intentions in the East became an increasing point of concern, it was vital to keep the U.S. Navy stationed in the Pacific. Control of the Atlantic Ocean could be left to the might of Britain's Royal Navy. For its part, the United States was becoming increasingly uncertain about Britain's will to stand up against the Nazis. Even those Americans who were inclined to come to Britain's aid feared that Britain might collapse, leaving its new ally high and dry. If the British Fleet were to fall into German hands, America would find itself exposed on two fronts several thousand miles apart. To Roosevelt the interests of Britain and the U.S. seemed inextricably joined in the Atlantic Ocean.

Yet there were numerous powerful voices in America who took a fundamentally different view. In the run-up to the European War in the 1930s, most Americans had felt secure in a country protected by those oceans and whose vast resources had not long ago helped to bring about the downfall of an earlier aggressive German regime. Roosevelt himself observed that he was "up against a public psychology of long-standing—psychology which comes very close to saying 'peace at any price.'"[6] And this public psychology had legislative backing in the Neutrality Acts. So entrenched was this spirit of isolationism that the country maintained a standing army of just a quarter of a million men. Even as German rearmament began to pick up momentum, America remained determined not to allow itself to be dragged into another European war. Roosevelt, however, was convinced that a strong Britain was America's first line of defense.

THREE

The Tizard Committee

THE RAF'S 1934 MOCK ATTACK ON LONDON had raised concerns about British readiness for a modern air war. Among the first voices to call for government action was Frederick Lindemann, whose letter to the *Times* on August 8 of that year was published under the heading "Science and Air Bombing."

> In the debate in the House of Commons on Monday on the proposed expansion of our air forces, it seemed to be taken for granted on all sides that there is, and can be, no defence against bombing aeroplanes and that we must rely entirely upon counter-attack and reprisals. That there is at present no means of preventing hostile bombers from depositing their loads of explosives, incendiary materials, gases, or bacteria upon their objectives I believe to be true; that no method can be devised to safeguard great centres of population from such a fate appears to me to be profoundly improbable. If no protective contrivance can be found and we are reduced to a policy of reprisals, the temptation to be "quickest on the draw" will be tremendous. . . . To adopt a defeatist attitude in the

> face of such a threat is inexcusable until it has definitely been shown that all the resources of science and invention have been exhausted. The problem is far too important and too urgent to be left to the casual endeavours of individuals or departments. The whole weight and influence of the government should be thrown into the scale to endeavour to find a solution.

Lindemann was closely associated with Winston Churchill and for much of the 1930s he too found himself on the political sidelines. Nevertheless moves were already afoot to do something about the situation. The RAF had formed a group of senior officers led by Air Marshal Sir Robert Brooke-Popham, whose brief was to improve the air defenses of London. They would be formed as a special subcommittee of the grandly titled Committee of Imperial Defence. But in Whitehall, the civil service proposed a far more radical response: the Air Ministry decided that the answer might well lie with civilian scientists rather than airmen.

The men who set the ball rolling were H. E. Wimperis, ten years in the post of director of scientific research at the Air Ministry, and his assistant A. P. Rowe. During the air exercises that summer Rowe had watched a demonstration of the only early-warning system in development, a prototype concrete "sound mirror" 200 feet long and 25 feet high facing Europe across the English Channel with the aim of picking up and concentrating the engine noise of incoming planes. It was readily apparent to Rowe that with planes getting faster and faster, an acoustic warning system would never work. It simply could not give enough warning to get British fighters scrambled into action in time. In any event it could be (and was) "jammed" by the noise of a passing milk cart. So, in June 1934, with no workable solution to the urgent early-warning problem, Rowe began a thorough trawl of all the Air Ministry files on air defense to see if there were any other proposals that might be worth pursuing. It was a chastening experience. After going through all fifty-three files he told Wimperis there were no useful ideas in any of them. He wrote that, "little or no effort had been made to call on science to find a way out" and "unless science evolves some new method of aiding air defence, we [would be] likely to lose the war if it started within ten years."[1]

Part of Wimperis's job at the Air Ministry was to evaluate the steady stream of claims to have invented a "death ray," a means of incapacitating men or machines by the projection of some kind of energy. Such was the parlous position in which Britain found itself that this possibility was now being canvassed at the highest level. But despite the generous offer of £1,000 to anyone who could successfully demonstrate such a ray, no one had come near to claiming the prize. On October 15, 1934, Wimperis sat down to lunch in the Athenaeum, one of London's most exclusive clubs. His guest was Archibald V. Hill, an eminent physiologist from University College. During World War I Hill had been the director of the Anti-Aircraft Experimental Section in the Army's Munitions Invention Department, where he had helped to create the new discipline that would later come to be known as operational research, the use of mathematical modeling and statistical analysis to improve operational decision-making. The subject of their discussion was "the use of radiant energy as a means of anti-aircraft defence." It was a momentous meeting, even though what it produced was not a death ray but a memorandum.

Inside a month a lengthy note from Wimperis landed on the desk of Lord Londonderry, the Secretary of State for Air. Acknowledging the special threat posed by hostile aircraft, the memo advocated that the British "intensify our research for defence measures and no avenue, however seemingly fantastic, must be left unexplored." Wimperis set out the need for a committee of scientists to tackle the problem of air defense, outlining some cutting-edge areas of research the committee might explore. He began with a prescient analysis of the scientific breakthroughs the next fifty years might hold, including television and "new ways of deriving our food from the soil." He predicted that we could also expect "the transmission by radiation of large amounts of electrical energy along clearly directed channels," which would inevitably be used "for the purposes of war."

Such research was at the leading edge of physical science, and Wimperis proposed that this was a job for a handful of the country's top scientists: "An excellent chairman might be found in Mr. Tizard, the present Chairman of Aeronautical Research Committee and a former R.F.C. pilot. The other members should, I suggest, be the

Professor A. V. Hill, F.R.S., and Professor [Patrick] Blackett, F.R.S., who was a Naval Officer before and during the War, and has since proved himself by his work at Cambridge as one of the best of the younger scientific leaders of the day."[2]

The scope of the committee, he proposed, should be wide enough to cover all possible developments. This was revolutionary stuff in 1934—the idea that the development of the country's defenses could be put in the hands of scientists rather than military men. In the past the role of scientists had been simply to try to improve on something that was already in existence. Yet now they were being asked to take on something of a totally different order, trying to find some completely new weapon or system that would radically alter the way the military men would fight their war. Unsurprisingly, Wimperis's committee would prove to be unpopular in some quarters.

Lord Londonderry, however, was far-sighted enough to appreciate the need for radical change. While great strides were being made in aeronautic design and the power units for aircraft, armament research remained in the hands of RAF officers intent on polishing what they already had. Londonderry accepted Wimperis's stark assessment of the situation and agreed to bring together a small committee to advise him at the Air Ministry. Its first members were exactly as Wimperis proposed: Tizard at the head, Hill, and Blackett—each of them a civilian scientist. Wimperis would be the only member from the Air Ministry, and Rowe would be the committee's secretary. Here was the genesis of a revolution in scientific warfare. Tizard was now the man charged with finding a way to stop the bombers.[3]

The new Committee for the Scientific Study of Air Defence (CSSAD) would eventually come to be known simply as the Tizard Committee. Tizard, though, had not exactly leapt at the chance to become chairman. The committee had been given a solely advisory capacity, and Tizard knew how difficult it would be to drive radical solutions through the intransigence of the civil service and the military. It would have no executive power, no money of its own, and no staff—not even a typist to deal with correspondence, which Tizard had to handle from his home address. Tizard, though, was keenly aware that this independence would give the committee far more innovating power than if it were tied to a particular branch of the civil service or armed forces. Indeed, before accepting the position Tizard wrote to Hill and Blackett suggesting that they would all "be in a stronger position if [they] were not paid anything like a salary or retaining fee."[4] Hill and Blackett agreed and these three scientists, unpaid and with no direct power, planned to start work in the new year. They were beginning a process that would transform first the air defenses of Britain, and eventually the nature of warfare on land, sea, and air across the globe.

Before the committee even sat for the first time Wimperis had taken what would prove to be a crucial step. Sometime during the first two weeks of 1935 he put in a phone call to the National Physical Laboratory at Slough, just outside London. Eager to get the new committee started on bold scientific projects, he wanted to talk to a man named Robert Watson-Watt, who, some years earlier, had started researching radio waves in connection with his work at the Meteorological Office.

The two men met at the Air Ministry on January 18, and Wimperis requested Watson-Watt's advice "on the practicability of proposals of the type colloquially called 'Death Ray,' that is, proposals for producing structural damage or functional derangement in enemy aircraft or their crews." Watson-Watt returned to Slough where he asked his assistant, Arnold Wilkins, to calculate how much power would have to be reradiated from a radio transmitter to heat a certain amount of water to a given temperature at a given distance. Such work was of course top-secret, but it didn't take Wilkins long to work out what was behind the request: "I noticed that the amount of water was just about the amount of blood in a man's body and that the temperature given was about fever temperature, so it seemed very likely that what was wanted was a 'Death Ray.' The power required was, of course fantastically large, and when I took the answer along, it was obvious that there was no chance of the Death Ray being produced by those means. Watson-Watt was not at all surprised. 'Well', he said, 'I wonder what we can do to help them.'"[5]

Watson-Watt had promised Wimperis that he would have an answer for him before the first meeting of the CSSAD on January 28. Clearly the death ray was a nonstarter but, according to Wilkins, Watson-Watt had other ideas. Working on problems of weather forecasting at the National Physical Laboratory, Watson-Watt had been bouncing pulsed radio waves off the ionosphere, a reflecting layer of the upper atmosphere about sixty miles up that, under suitable conditions, can act as a perfect mirror for radio waves. The elapsed time would give the height of the ionosphere to a very high degree of accuracy. The method they were using had been invented in America some ten years earlier by Gregory Breit and Merle Tuve, and was in regular research use around the world. No one, however, had developed the idea of applying it to the detection of aircraft in the way now suggested by Watson-Watt. Wilkins recalled his contribution: "To find out whether it was possible to turn what was known to practical use Watson-Watt asked me to find out what power would be required to produce a detectable signal from an aircraft at such and such a range. I then did two things. I assumed that the aircraft would have the re-radiating properties of a half-wave aerial, and I took it that the plane would measure about 25 m horizontally and 3 1/2 m vertically."[6]

Wilkins's calculations showed that the reflections from such a tiny object as a distant airplane would be about 10^{-19} the strength of that which was sent out. Nonetheless, he worked out that with a big increase in the transmitted power and sufficiently sensitive receiving apparatus, it should be possible to pick up the very weak signal that would be reflected by a plane many miles distant from the transmitter. Wilkins had sketched out the theoretical basis for the location of aircraft by the use of radio waves. Both men immediately saw that the possibilities were astounding. There was the germ here of a new way of seeing the bombers coming.

Previously the detection of incoming aircraft had been limited to the visual identification of the Observer Corps. As Rowe had seen, the experimental sound mirrors were handicapped: with sound traveling at around 650 mph at sea level, and the bomber itself doing over 200 miles an hour, the amount of extra warning was limited. Radio waves however, like light, travel at 186,000 miles per *second*, and the potential range of detection is also vastly increased. So this new technology held enormous potential. That potential was clearly seen and understood by Tizard and the committee at their first meeting at the end of January, and they asked Robert Watson-Watt to outline the theory in writing. It took him just a fortnight to produce the paper called "Detection and Location Of Aircraft By Radio Methods," and on February 15, Wimperis approached Air Marshal Sir Hugh Dowding with a request for £10,000 to fund initial research on the system.

In simple terms, radio waves would be emitted by a transmitter. They would scatter off a solid object and a tiny percentage of them would bounce back in the same direction to a receiver, like an echo. If the object were moving, there would also be a slight change in the frequency of the returned waves. Because the waves are traveling at the speed of light, a simple calculation multiplying the time it takes for the echo to travel out and back by the speed of light itself, and then dividing by two (because the signal travels out and back) will give you the distance of that object from the transmitting and receiving station. Not to mention the crucial information that the solid object (possibly an enemy bomber heading your way) is there in the first place.

Watson-Watt proposed a pulsed echo system in which radio waves are transmitted in very short pulses. When you're working at the speed of light, the pulses you send out must be incredibly short in order not to drown out the extremely faint return echo from a small plane that may be a hundred miles away. In fact you must be able to generate pulses measured in microseconds, with (comparatively) huge gaps between them—yawning silences lasting many milliseconds. The general effect can be replicated by walking toward a sheer cliff with a whistle. Every time you give a good blast on the whistle the echo will return a few moments later. The closer you get, the shorter the delay. And, sooner or later, you'll find the echo coming back before you have finished whistling. Your transmission is drowning out its returning echo. But stop right there, shorten your whistle, and once again you can distinguish the returning sound. The shorter your whistle, the closer you can get. Conversely, the longer the gap you leave between whistles, the farther you can be from your target, because your next pulse doesn't overlap the incoming distant echo. Therefore, there are minimum and maximum ranges inherent in the pulsed system—a major issue if you're sitting in a night fighter trying to home in on an enemy bomber in total darkness.

Dowding, however, wanted a practical test carried out, so the committee arranged for a test flight to take place on February 26, 1935, in which a twin-engine Heyford biplane would fly past the BBC's radio transmitters at Daventry in the English Midlands. On the ground Wilkins, Watson-Watt, and Rowe stared nervously at the hastily rigged cathode-ray tube display as the plane approached. They were elated as they saw a small green blob gradually rise and fall as the plane flew up and past. They were able to correctly calculate the plane was some eight miles distant. When the news reached London, Air Marshal Dowding told Wimperis he could have all the money he wanted, "within reason."[7] RDF, or radio direction finding, was on its way.

But there was a dark cloud on the horizon. One name conspicuous by its absence on the committee's roster was that of Lindemann. While Tizard wanted to keep his scientific committee free of political influence, Lindemann was happy to use his political influence in pursuit of his scientific aims. Along with Churchill he had spent years agitating for the establishment of a powerful political committee to look into the problems of air defense. Less than three weeks after Tizard's advisory committee first met, Lindemann made a passionate speech advocating this course to an influential committee of Conservative backbench MPs. The ailing Socialist Prime Minister, Ramsay MacDonald, heavily dependent on Conservative support, finally caved in and agreed; a committee was to be set up under the man shortly to become Air Minister, Lord Swinton. Among its members would be Lindemann. He brought to it a passion born of a stark understanding of what might be achieved by the bomber in a modern war. His vision of those possibilities, and the urgency he attached to the need to find a response come through clearly in the notes he made in preparing that speech: "There is no time for the League of Nations to function, no time for democracy to make its voice heard. It may well be that the result of the war will be decided within the first six hours."[8]

The task of Swinton's Air Defence Research Committee (ADRC) was to consider the problems of air defense in the round, which is to say it had to consider all the political and economic issues of which the scientific possibilities were merely a part. The ADRC would report directly to the Committee for Imperial Defence, the body with supreme oversight of military matters. Tizard's CSSAD was merely advisory; this new committee would have executive powers, and it would oversee the work of the CSSAD. Though Tizard himself was to have a seat on Swinton's committee, the prime minister made it clear that the new committee was to have "the direction and control of the whole inquiry."[9]

As if this situation were not sufficiently confusing, in June 1935 Prime Minister MacDonald lost power and Stanley Baldwin, a Conservative, took his place. Churchill had been making waves from the backbenches for years about the lack of urgency in developing Britain's air defenses. Now, with the change of government, Lord Swinton decided that it was better to have such a maverick voice in rather than out and invited Churchill to join his committee, the ADRC. Churchill agreed, but with one stipulation—that Lindemann, his key technical advisor, already serving on ADRC, should now be drafted onto its technical subcommittee, the CSSAD—Tizard's committee.

Now, it seemed, the Tizard Committee had a cuckoo in the nest. For though Lindemann was passionately concerned about air defense and the need to bring new scientific solutions to the table, his favored path was one that fundamentally differed from that which Tizard and the rest of the committee had already embarked upon. Once appointed to the committee Lindemann arrived with all the zeal of the freshman. And he had ideas of his own. He began to set them out in a letter to Tizard dated July 1, 1935. Radio direction finding was not among them.

Instead he advocated the development of a new technology. As a trained statistician he had worked out the odds against an anti-aircraft gun on the ground being able to hit the moving target of a high-speed bomber. At the time, it was virtually impossible for a gunner to know, as he pressed the fire-button, where the aircraft flying thousands of feet above him would be by the time his shell arrived. Even if he could perform that complex calculation, it would only require a slight deviation of course for the plane to miss the target entirely. Lindemann's conclusion? That the only way to bring down an aircraft by antiaircraft fire was to put up a barrage of shells that would simply fill a whole area of the sky. That, according to his calculations, could take as many as 90,000 shells to bring down a single plane, clearly an absurd number. Hence his search for a new technology—aerial mines.

The idea was that raiding airplanes should run into a curtain of tiny explosive devices each weighing only about four or five ounces (100 grams) and suspended by a thin line or cable from a small parachute. The small bombs would be dropped by a plane, or possibly fired from a gun or by rocket, into the path of the oncoming raider. Again Lindemann the statistician had been at work. "Supposing [each mine] to be of the order of 100 grams and supposing it to hang upon a wire 10 metres long, then with aeroplanes of roughly 30 metres span one would require a thousand such mines to bring down 1/3 of the aeroplanes passing through a region 2 kilometres broad and half a kilometre deep."[10]

Outlandish as it may seem now to try to bring down bombers by filling the space they were about to fly into with thousands of slowly descending mini-bombs, Lindemann was not the only advocate of the aerial mine. Perhaps from the perspective of 1935 it would have seemed no more outlandish than the idea of detecting planes at distances of over 100 miles with radio waves. Nevertheless this idea of Lindemann's found little favor with the other members of the Tizard Committee, and neither did his support for researching the use of infrared rays as a means of aircraft location. Nor did the suggestion that attempts should be made to find some kind of substance that could be propelled into the path of incoming aircraft causing their engines to explode or fail. In short, Lindemann's ideas fell on stony ground. But he stressed the importance of diverting resources to his ideas as soon as possible: "It is admitted that . . . these suggestions would require a great deal of work before they could become practical proposals. In view of the importance of the problem, however, and the limitation of sound location, no effort should be spared to bring them to fruition."[11]

The next year, up through the summer of 1936, would be a fruitful period for the Tizard Committee and the early development of radar.* Yet at the same time it was a year of bitter clashes over the direction the committee should take, clashes that would become ever more personal between Lindemann and Tizard as the year unfolded.

The seeds of the bitterness had been sown even before Lindemann joined CSSAD. Tizard had taken exception to criticisms Lindemann had apparently made of his committee's work to the overseeing Committee of Imperial Defence. Now, at each of the CSSAD's regular meetings Lindemann would continue his campaign for the diversion of more resources into research in the use of aerial mines. He made no secret of his view that the committee was wasting time and accomplishing little, apart from the progress being made on radio direction finding. On one key aspect of the matter Lindemann was absolutely right, for Tizard had already come to the firm conclusion that the successful defense of the country might well rest upon the development of this untried technology. It was already making far greater demands on the resources of the committee than anything else, leaving little room for pursuing Lindemann's ideas. And Tizard's faith was already being rewarded with substantial progress.

After receiving Watson-Watt's paper, Tizard had arranged for practical research to begin on a bleakly beautiful stretch of English coastline facing Continental Europe. Orfordness, in Suffolk, was well known to Tizard from the years he had spent in aeronautical research during World War I. On a sparsely populated stretch of the east coast, it had the unique advantage of being remote and deserted yet within relatively easy reach of London. It was here that Robert Watson-Watt and a small team of scientists from the National Physical Laboratory, under the cover of doing ionospheric research, were already starting to construct the first prototype military radar. Day-to-day operations were overseen by Arnold Wilkins, along with another scientist from the team at the National Physical Laboratory,

*The term "radar" will be used throughout as the generic term for radio wave detection systems. Electronic engineers working for the U.S. Navy in the early 1940s coined the term, an acronym from "RAdio Detection And Ranging.

a man with the exotic name of Labouchere Hillyer Bainbridge-Bell. Their working environment was primitive, their laboratories little more than sheds. This was a place for the hardy and resilient, and Watson-Watt and Tizard were quick to see that they would need the brightest young minds in physics if they were to achieve anything in these conditions before war came: men like twenty-four-year-old Edward "Taffy" Bowen, a Welshman from Swansea. Bowen was the final founding member of the elite band of radar scientists tasked with inventing a defense to the bomber, and Watson-Watt had already marked him out for great things.

On June 17, 1935, the fledgling team scored their first success when they were able to detect a Supermarine Scapa flying boat passing Orfordness at a distance of seventeen miles. It was a significant milestone and convinced Tizard, and his committee, that they were very much on the right track.

FOUR

Bawdsey Manor

In the early days of CSSAD at Orfordness, the purpose of radar was quite clear. The task was to construct an early warning system that could be used to alert Britain's fighter defenses that German bombers were on their way. If the RDF system could pick up an incoming bomber a hundred miles from the coast then Fighter Command would have twenty or thirty minutes to get fighters into the air and to an altitude at which they were ideally placed to attack the enemy aircraft. Predicting exactly where the aircraft would arrive was of course another matter, but systems were very quickly developed to track the progress of incoming attackers so as to plot their height and direction. At the suggestion of A. P. Rowe, RDF came to stand for "range and direction finding"; it would keep the same acronym as the already well-known radio direction finding, and therefore disguise its increased ability to track a target's range as well. All of the team's energy was thrown into constructing a functional RDF-based warning system and field-testing prototypes as quickly as they were developed. In December 1935 preliminary plans were laid to build five radio towers that would protect the flight-path approach to London.

Soon, however, Robert Watson-Watt's team began to outgrow the facilities at Orfordness. A bigger, more permanent location had to be found, preferably without losing those advantages afforded by being close to the Suffolk coastline. The answer came with the availability of a remote English country house just up the coast. A greater contrast to the bleak marshland of Orfordness would be hard to imagine. Their new home perched on a small cliff above a sandy beach. It had spacious lawns, peach trees, and bougainvillea. This was Bawdsey Manor, an imposing Victorian Gothic mansion built at the end of the nineteenth century by local MP William Cuthbert Quilter. Over twenty years he had developed the house by adding further towers and façades in Flemish, Tudor, Jacobean, French chateau, and Oriental styles. Through the twenties and thirties it was home to his large family and the scene of lavish house parties. When the scientists arrived the billiard table was about to be thrown out. Eddie Bowen bought it for £25 and it stayed where it was, entertainment at the end of a demanding day. With its stables and outbuildings converted to workshops, the Bawdsey Research Station was to be the nursery of radar until the outbreak of war.

The move to Bawdsey took place in March 1936 and coincided with another far more momentous move as Hitler's army marched unopposed into the Rhineland in direct contravention of the demilitarization of the region imposed by the Treaty of Versailles. This was perhaps Hitler's greatest gamble throughout the 1930s. His most senior generals were advising against it. At that early stage the Wehrmacht would have been no match for the French Army, yet the French government, facing imminent elections, decided not to resist the move. Hitler had correctly divined that there was no appetite in France or any other of the Western powers for another military conflict, and that many politicians in France and Great Britain in particular felt that the provisions of the Versailles Treaty were far too stringent and that such a conflict would have scant moral justification. In the event, of course, it served only to strengthen Hitler's belief in his own ability to pull off an audacious coup. Hitler himself would later write: "The forty-eight hours after the march into the Rhineland were the most nerve-racking in my life. If the French had then marched into the Rhineland we would have had to withdraw with our tails between our legs, for the military resources at our disposal would have been wholly inadequate for even a moderate resistance."[1]

An opportunity to take on Hitler had been missed, but the three-year delay before the outbreak of war was to be put to good use by the radar scientists moving into Bawdsey. One man who might, perhaps, have picked up the gauntlet thrown down by Hitler with the remilitarization of the Rhineland was Winston Churchill. Churchill had served as an army officer in South Africa during the Boer War, and he had spent many years heading the navy as First Lord of the Admiralty. His knowledge of air warfare, however, was scant by comparison. But by joining Swinton's committee in 1935 he was now at the heart of air policy, well positioned to absorb the latest research and information on the subject. From that summer onward he became acutely aware not only of the weakness of Britain's existing air defenses, but also of the encouraging progress being made on the development of radar under the auspices of Henry Tizard's subcommittee. As ever, though, Frederick Lindemann strongly influenced Churchill's opinions on technological issues; indeed for some months Lindemann was staying with Churchill at Chartwell, his country house in Kent. As the rancor between Lindemann and Tizard grew, so did Lindemann's vocal enthusiasm for aerial mines and his skepticism that RDF would ever amount to anything.

In June Churchill put pen to paper to complain to Lord Swinton about the way the Tizard committee was being run and about Swinton's apparent support for Tizard. He echoed Lindemann's view that members of Tizard's committee were dragging their feet, and he complained about their reluctance to divert resources into Lindemann's pet project, aerial mines.

> 22nd June 1936. Confidential.
> The differences upon the Scientific Committee are not as you suggest of a technical or abstruse character. They are differences about the method and procedure to be used in testing certain ideas, which if found sound would open a new domain to anti-aircraft artillery, as well as helping in other ways known to you. The experiments are neither large nor expensive, but they must be numerous, and can only advance by repeated trial and error.

Churchill also took exception to a letter Tizard had sent Lindemann in defence of the committee's position. "His first step has been to write a very offensive letter to Lindemann which I should have thought would make, and was perhaps intended to make, their future relations impossible."[2]

It was part of an increasingly acerbic exchange the two scientists had been having throughout that June. Tizard had written early in the month to set out his unhappiness at the way Lindemann was pressing his case. It prompted a vitriolic reply from Lindemann handwritten on June 25, 1936, on notepaper from Claridge's (top hotels were Lindemann's natural habitat). The current rate of progress, he wrote, would be fine if there were "ten or fifteen" years to prepare for war, instead of, as he believed, mere months. He was forced to admit that Watson-Watt's work at Bawdsey was going well, but warned Tizard that he was prepared to use any influence he had to speed things up: "In view of the immense importance of the question [of air defense] and holding the views I do as to its urgency, you will not be surprised that I have used every means at my disposal. I am sorry if this offends you, but the matter is too vital to justify one in refraining from action in order to salve anybody's amour propre."[3]

Tizard's reply on July 5 was equally blunt: "I got your letter of the 25th June in reply to mine. You need not worry about salving my amour propre. I haven't got any. My quarrel with you was not that my dignity has been affronted, but that your way of getting on with the job is the wrong one, that far from 'accelerating progress' you are retarding it." Tizard was angry that Lindemann had been "complaining to other people that [the committee] were slackers," and finished his letter with a polite threat: "If you persist in the attitude disclosed by your letter I do not think that we can remain members of the same Committee; but do give co-operation a further trial. I am much more interested in defeating the enemy than in defeating you! Yours sincerely, H T Tizard."[4]

Things could not go on like this, and with Churchill pressing his case in the parent committee, Lindemann was not going to back down. Within a couple of weeks two letters dropped onto the desk of Lord Swinton, the Air Minister and head of ADRC. They were the resignations of Blackett and Hill, the two scientists originally recruited to work alongside Tizard. Relations on the committee had clearly gone beyond frosty and now, with Hitler's armies on the march and war breaking out in Spain, the committee charged with finding a scientific solution to the threat of the bomber had simply disintegrated. Lord Swinton was about to earn his considerable salary.

There are times when the British obsession with committees could produce confusion and animosity. But there are times when the politics of pragmatism can find a simple answer to an otherwise intractable problem. On September 3, 1936, Lord Swinton wrote to Lindemann to inform him that he was dissolving the committee after receiving the resignations of Hill, Blackett, and finally Tizard himself. Lindemann replied, acknowledging that the resignations were due to "irritation because I endeavoured to accelerate the very slow progress which has been made in the investigations."[5] Once again he set out his case—that more effort should have been put into aerial mines and infrared detection. It seems to have cut little ice with Swinton, who had already decided on his course of action. In early November he sent a handwritten note to Lindemann explaining that he had decided to reconstitute the committee under Tizard, but without Lindemann: "I found that I could not secure the co-operation of the other members on the basis of the old membership; and my colleagues and I regarded the work of the committee as much too valuable to be allowed to lapse."[6]

The committee had in fact been reformed, and had already met in early October, with all the old membership minus Lindemann but with the addition of Cambridge professor Edward Appleton, Britain's leading expert on radio waves. This was a clear endorsement by Swinton of Tizard's chairmanship of the committee and the direction it had taken. RDF was the priority, and Tizard the victor—for the time being.

In part, Swinton had been convinced to preserve CSSAD and encourage its RDF work by events that had started during the summer of 1936 at the small grass-runway airfield at Biggin Hill in Kent. The experimental work at Orfordness and Bawdsey Manor held out enormous potential. Watson-Watt and his team had built prototypes that clearly established the possibility of detecting incoming aircraft at distances as great as seventy-five miles, far beyond what could be achieved by visual observation or sound locators. Yet this was a long way from promising a comprehensive system of defense. An incoming bomber could cover that seventy-five miles in close to twenty minutes. To be able to successfully engage that bomber meant getting a fighter into the air and then directing it to the precise point where it could intersect with the bomber's flight path—assuming the bomber maintained a steady course.

In these pre-radar days, a military aircraft was a dangerous and often temperamental thing; its pilot almost invariably a young man trying to make split-second, and independent, decisions. Yet what was required in order to make proper use of the information which the new radar devices could deliver was a system where some technical operator, his feet safely on the ground, could co-ordinate a number of airplanes, instructing each pilot what to do. And this in the heat of battle when the pilot's very life was under threat. To devise such a system was a tall order. To get it accepted by the pilots seemed virtually impossible. This Tizard understood only too well—he himself was a pilot.

At the outbreak of World War I Tizard was already twenty-nine years old and a successful scientist teaching at Oxford University. He had enlisted in the early days of the war with the Royal Garrison Artillery, where he quickly became an officer training recruits in the use of a fairly primitive anti-aircraft gun. He was impressed by neither the gun nor the methods in the training manual. "One might possibly have hit a low-flying Zeppelin with it," he wrote, "but the chance of hitting an aeroplane was negligible." This experience taught him both something of the difficulties of defending against air attack, and a healthy skepticism of military authority. Soon, however, it was felt that his scientific skills could be put to better use in the new field of aeronautics and he was invited to transfer in July 1915 to the Royal Flying Corps "on probation for an appointment as an assistant Equipment Officer in connection with experiments." He was attached to the Central Flying School at Upavon in Wiltshire, where the opportunities for the scientific development of aerial warfare were exciting. Heavier-than-air flight itself was, after all, only some twelve years old at this point. He quickly began work on the problems of bomb-aiming and devised a photographic system for measuring the time lag on the fall of bombs from specific heights. It didn't take him long to work out, though, that simple flying errors were much more frequently the cause of missing targets than miscalculation on time lag. "I then came to the conclusion," he wrote, "that it would be very difficult for any scientific man to do really important relevant work unless he himself learned to fly."[7] It seemed logical to Tizard, but the War Office objected. There was a war on, and the job of the Central Flying School was to train pilots for operational squadrons. Such was the attrition rate for pilots at the front that there was a pressing need for a constant supply of new ones. Training a pilot who would never fly in combat seemed to the War Office something of an impediment to the central task of the school. But Tizard was not to be brushed off so easily and pressed the point with his commanding officer. This time the War Office came up with a solution, one they might perhaps have felt that Tizard would find unacceptable. They agreed that he could be taught to fly, but only on days when the weather was too bad for the trainee pilots to be taken up. Characteristically, Tizard jumped at the chance, commenting that it was good because one got used to the difficulties right away.

It was something of a baptism of fire, but a vital element in Tizard's lifelong ability to understand and appreciate what each new scientific development meant to the men at the sharp end, the pilots. Already, at Upavon, a relatively aged thirty years old and a scientist and academic, he knew there was a potential gulf of understanding between him and the young men who were being trained to fly their frail machines into battle. That Tizard was prepared to learn to fly in the most challenging conditions, and thereafter was always willing to put his own ideas into practice in the cockpit, would help him to build bridges and smooth the path for the operational application of his ideas. This flight training, and the rapport he was able to develop with airmen, was to stand him in very good stead some twenty years on when it came to the critical developments at Biggin Hill.

The answer to the problem of how to convert simple detection of an incoming bomber into a successful engagement, Tizard knew, lay in real-time experimentation. A technique had to be developed as quickly as possible in the relative calm of the troubled peace that now lay across Europe. It would be worse than useless to try to forge such a system in the heat of battle. By 1936, Tizard had for seven years been Rector of Imperial College in London, one of Britain's foremost seats of scientific learning. Though now a civilian, his time as an Air Force officer and a pilot gave him an edge when it came to securing the cooperation of the RAF. By August of that year he had managed to persuade Air Marshal Hugh Dowding, the newly appointed head of Fighter Command, to allocate a squadron of aircraft based at the Biggin Hill airfield for a series of experiments set to last for two months. The Gloster Gauntlets of No. 32 Squadron were the fighters assigned the task of intercepting three Hawker Hinds that would simulate the incoming enemy bombers—all, of course, under control from the ground.

At first, the Biggin Hill experiments were conducted under such tight security that the pilots did not even know of the existence of radar. In fact none of the officers at Biggin Hill had any idea of the object of the experiment until Tizard arrived. He then explained to them the purpose of the exercise in the most guarded of terms. He told them that a ground station would be able to give him approximately fifteen minutes' warning of the approach of the Hawker "bombers." He would have rough details of their height, speed, and course, and this information would be refreshed every five minutes. The task of the team at Biggin Hill would be to work out how they would plan fighter interceptions using this information. They began by setting up five trial interceptions each day. As soon as the Hawker Hinds were spotted by radar, their course and speed were calculated to produce a possible interception point. The No. 32 Squadron would then be ordered into the air with instructions on their course and the height to which they should climb. Every five minutes they would receive fresh, updated instructions over their radio. Within a few weeks the team was achieving nearly 100 percent success, as fighters were guided close enough to the incoming bombers to visually spot and engage them. With the satisfactory results Tizard revealed to his team how the course of the "bombers" was being obtained by the reflection of radio waves.

Now that the RDF cat was out of the bag, Tizard told them that the information on the bombers' position and course would now come not merely at five-minute intervals but in a constant stream. The fighters would be directed almost entirely remotely by controllers on the ground. The job of the pilots would be to follow these instructions until they got to the appropriate position in time for the bombers to arrive. But now the difficulty was stepped up still further. Until now the approaching bombers had maintained a constant course, but what if they were to take evasive action or simply to change course toward a different target? Once the experiments had started Tizard was quickly able to work out a simple method of tackling the problem that they called the principle of equal angles. As soon as the bomber was observed to change direction the ground controller simply drew a straight line on his plot between the bomber's new position and the position of the intercepting fighter. This would be the baseline of an isosceles triangle. The bomber's new course would form the second line of the triangle, and after a few simple calculations on the speed of the aircraft, the third line could be drawn—the fighter's interception course. If the bomber changed course again, the calculations were simply repeated. So it was that, at Biggin Hill in the autumn of 1936, the basic principles of radar-controlled defense were worked out for the first time.

The experiments would carry on well into the new year, but by the end of 1936, radar had so proved its potential that the Air Ministry was prepared to commit the air defense of the country to this new technology. That commitment would require huge resources, and they were forthcoming. In February 1937, a training school for RAF personnel was opened at Bawdsey. If radar were to become a functional defense system then it would have to be operated by servicemen, not scientists. Fortunately the trainees took to it almost effortlessly. Alongside the school, workmen were busy constructing the first operational radar station, and two months later, in May, it was handed over to the RAF and their first trainees. This was the first of a chain of stations whose two- and three-hundred-foot-high towers would eventually stretch the length of Great Britain, from Ventnor on the Isle of Wight north to the Firth of Tay in Scotland.

By August 1937, when some important defense exercises were due to be held, three stations were up and running. For the first time the RAF grandees would be able to observe radar in action on an operational scale. In fact the system performed disastrously, with the same aircraft being plotted in different positions at the same time by different radar stations. Nevertheless, the exercises had usefully revealed a flaw that could be fixed. The solution to the problem was a central filter room, where a single team would coordinate and refine the incoming data from different stations, and it would turn out to be a key element of the system. This minor hiccup did nothing to dent the enthusiasm of the Air Ministry for the construction of what came to be known as "Chain Home," and stations began to go up at a rapid pace.

The early years at Orfordness and Bawdsey were tremendously fertile, when new possibilities and techniques for utilizing this new technology seemed to tumble out one after another. As Eddie Bowen recalled:

> It became clear that it was Tizard and the members of his Committee who were thinking ahead and providing the leads. Tizard was confident that the air-warning range of nearly 100 miles which had already been achieved by the chain system, when married to the filter and interception procedures which came out of the Biggin Hill experiment, had an excellent chance of repelling the daylight attack on Britain. The German Air Force, he argued, would then turn to night bombing. What could be done about this? This was how Tizard's mind worked and he was to provide many examples of a similar kind over the years. His prediction that the German Air Force would be beaten back in daylight, and that it would then turn to night bombing must rank as one of the best examples of technological forecasting made in the twentieth century.[8]

Tizard was convinced that what was needed to counter bombers arriving at night or in poor weather was airborne radar that allowed pilots to intercept the enemy without visual contact. Airborne radar was high on his list of projects to be tackled, but someone with enthusiasm for the task needed to take it on. That man was Eddie Bowen, but he would have to persuade his boss, the relatively conservative Watson-Watt, that this fanciful notion was worth diverting someone from the comparatively mundane task of putting the Chain Home system into action. This he managed to do over several pints of beer one Friday night in the Crown and Castle Hotel in Orford village. Such was the strain on their manpower at that time, though, that the new airborne radar group had a staff of one, Eddie Bowen. Even his job was part-time at first.

When Bowen began to draw up the specification for what an airborne radar might look like, the requirements seemed tricky indeed. First, it would have to weigh no more than two hundred pounds. Space in an airplane was very limited, and so was the available electrical power. The equipment would have to fit into a total of no more than eight cubic feet and the maximum power available on an aircraft at that time was a mere five hundred watts—the equivalent of a few light bulbs. It would have to withstand the plane's vibrations. In addition, all the emphasis in aircraft design was on increasing speed. Anything that would increase the aerodynamic drag—like a radar antenna, for example—would be heavily frowned upon. Which meant that Bowen would have to try to make the system work with short stub antennas no more than a foot long. As if all that were not sufficiently difficult—the equipment would have to be simple enough to be used by a pilot who had his hands full already; if it were to be installed in a two-man fighter, even with a fully trained radar operator, the system would still have to be straightforward enough to be used in cramped and noisy conditions, even in total darkness.

A daunting prospect, but such was the pace of events at Bawdsey at this time and the ingenuity of the small team that Bowen managed to gather around him, that within a year or so, in August 1937, an RAF Anson took off with a fully operational and independent airborne radar set on board operating on a wavelength of 1.25 meters. A second Anson was used as a target that day, but the real victory was when the two engineers on board were able to obtain clear echoes from ships off the coast at nearby Felixstowe at ranges of two to three miles. It was cumbersome and limited, but it was working. And it was airborne.

Three weeks later came a chance to put the equipment to a serious test. News came through of an air–sea exercise due to take place on September 4, 1937. The object was to assess the ability of Coastal Command aircraft to find ships in the North Sea. An unknown number of British naval vessels would sail through the Straits of Dover and north to Invergordon. They would follow a zigzag path in order to avoid detection. The job of the Coastal Command aircraft was to locate this naval task force in a huge area of open sea. This was too good an opportunity for Bowen and his men to miss—the chance to put the capabilities of their lone radar-equipped Anson to the test.

As dawn broke that morning the airfield was shrouded in mist. But it was not long before the morning fog began to lift, and as soon as they could see the trees at the end of the runway Bowen, Keith Wood, another radar engineer, and the only available pilot, Sergeant Naish, lifted off into the clearing sky. Naish, as it happened, was the perfect man to have on board. Newly commissioned into the Air Force, and previously a master mariner in the merchant marine, he knew the North Sea like the back of his hand. As they climbed up and out over the coastline they got a clear sight of the Coastal Command flying boats heading off to the north. Bowen and his Anson crew, however, had a different strategy in mind, and began to search the sea directly east of the Suffolk coast, flying a square pattern at 3,000 feet with the radar scanning ahead of and below them. Occasionally they could pick out the echoes of small craft on the sea beneath. Suddenly, there was a much larger echo—this time at a range of five or six miles. Turning to fly in the direction of the echo, they were soon able to make out most of the fleet—including the aircraft carrier *Courageous*, the cruiser *Southampton*, and a number of destroyers. Unfortunately, theirs was something of a freelance enterprise, and the fleet below were somewhat alarmed at the appearance of this strange, rogue aircraft homing in on them. Signal lights began flashing between the ships, followed immediately by warning gunfire. Then, as an unexpected bonus, the Anson's radar began to pick up the echoes of pursuit aircraft taking off from the deck of the carrier.

The carrier's full complement of fifteen Swordfish was taking to the skies. It was the first time Bowen and his crew had been able to detect aircraft between themselves and the sea, but it was no time to hang around, and so the Anson swung quickly away and began climbing to 9,000 feet. They had already demonstrated an extraordinary ability to find both ships and aircraft, easily beating Coastal Command at their own game. But they wanted more, and set off in pursuit of the battleship *Rodney*.

This turned out to be a dangerous choice. *Rodney* proved more elusive than the rest of the fleet, and by the time they decided to give up they were running short of fuel. The weather had begun to close in, so much so that the Coastal Command aircraft had already been forced to give up the chase. They had been searching for so long that they were also now uncertain of their position. Once again their radar was to prove its worth, though, as having climbed to 12,000 feet to get above the gathering cloud, their radar set was able to show them precisely when they crossed the coastline and could begin their eight-mile descent to the aerodrome. The first steps had been taken toward systems that would become both airborne interception (AI) and air to surface vessel (ASV) radar. Bowen recalled:

> We had found the Fleet under conditions which had grounded Coastal Command, we had detected other aircraft for the first time with a self-contained airborne radar and, simply by returning home in one piece, had demonstrated some of its navigational capabilities. From then on our path was much easier. A. P. Rowe was never a great admirer of the airborne group, but of this episode he was later moved to say 'This, had they known, was the writing on the wall for the German Submarine Service.' It was exactly one year and 364 days before the outbreak of war.[9]

But if this equipment was going to make any difference to the war it would need to be standardized and mass manufactured. Before it could be put into production an answer had to be found to a fundamental question—what was it *for*?

Airborne interception was the ultimate goal. But finding a lone attacker in the emptiness of the sky in heavy overcast, or possibly even at night, was pretty much the aerial equivalent of the search for a needle in the proverbial haystack. The new possibility of airborne radar held out some hope of success, but the needle was both elusive and very, very small. Would it not be helpful, then, if the target were something bigger, slower, and preferably in a position that was reasonably predictable to the operator of an airborne radar set? In fact Bowen's test flight in September 1937 had found just such a target in the ships of the Royal Navy task force. Ships, of course, have the virtue of being big, slow, and below. It made them, in many ways, the ideal target for the new tool of airborne radar, something which was not lost on the Commander in Chief of Coastal Command, Sir Philip Joubert de la Ferté, who was immediately enthusiastic about the potential of airborne radar in his command's job of patrolling the British coastal waters. He decided to travel out to Suffolk to see for himself what it could do for his men.

That visit would push the development of airborne radar for ASV use right to the top of the priority list for Bowen and the small team he would now have at his disposal throughout 1938. A successful demonstration flight with Joubert and Watson-Watt on board on July 4, 1938, gave the project added impetus, though Bowen thought he might have gone a bit far when his speculation that "one of these radars would soon be installed in the nose of every aircraft" produced profoundly skeptical looks. Conversely, for airborne interception, 1938 was something of a lost year. Fighter Command's Sir Hugh Dowding had also come to visit that summer. He, too, was impressed by the system's ability to find ships at sea. But with Germany's territorial demands growing by the month, Dowding was clearly doubtful that the obvious potential of airborne interception could be realized in time.

These were among the earliest in a series of demonstration flights over the next twelve months with some of the key decision makers of Britain's defenses on board. Churchill and Lindemann both came to Suffolk to see what had been achieved, although Churchill—being in his sixties, portly, and a busy politician—never actually left the ground. Lindemann did, but might have wished he hadn't, when smoke started pouring from the engine of the test plane he was flying in. These were promising developments but clearly not yet ready to be entrusted with the lives of airmen.

Dowding, however, was not a man to give up—particularly on something that might be an answer to one of Fighter Command's greatest fears, and he returned to Suffolk to monitor the progress Bowen's team was making on airborne interception. Their prototype airborne radar had the potential to put a plane within sight of its target, and, theoretically at least, it could do so at night, and in total darkness. Chain Home would give the fighters a head start in combating any daylight raid by German bombers, but, like Tizard, Dowding was now convinced that sooner or later the bombers would come by night, and that would make airborne radar an imperative. Dowding wanted to take another look.

The demonstration went well. The minimum range problem had been just about cracked. But Dowding seemed less than completely satisfied. Bowen and his team were puzzled. After the flight was over Dowding stayed on to talk to them. Reviewing the current capabilities of airborne radar at first hand had, it seemed, only served to highlight his worries about the night fighter problem, which he began to set out to them in clear and simple terms. The night fighter's task, he said, would be fundamentally different from that of the day fighter. The job would almost certainly be a long drawn-out one, and the aircraft would need exceptional range and high endurance. He foresaw a navigational problem, because in an extended interception attempt the night fighter pilot could end up well away from his base and unsure of his position (a problem that Bowen knew only too well). If a radar set were to be installed to help with target location or navigation, this would surely be demanding too much of the pilot, who would be faced with the seemingly incompatible tasks of looking at a relatively bright cathode ray tube while his eyes remained accustomed to the dark in order to make visual contact with a tiny, unlit, enemy plane in the surrounding darkness. Any radar-equipped fighter, he was sure, would have to be at least a two-seater to allow for a dedicated radar operator. The trouble was the Blenheim, the only appropriate two-seater available at that time, was a slow and cumbersome aircraft. The Blenheim had started life as a fast civilian machine, and had then been taken up by the military as a light bomber. Now it was being pressed into service as a fighter, a role in which it was clearly outclassed by the new generations of purpose-built fighters.

Another problem was the need to be sure that any aircraft engaged was, in fact, an enemy plane. The identifying roundels used to distinguish RAF planes would, of course, be useless at night. And, finally, Dowding concluded there was also a problem with firepower, because the night fighter would seldom get more than one chance to attack his target. A victim not brought down at the first attempt would simply dive and turn sharply away, and with the advantage of stealth gone even a radar-equipped fighter would find it virtually impossible to regain contact. Air interception by night was a vital and valuable prize, but as Dowding departed for London, Bowen was left thinking that his successful demonstration had merely thrown the whole problem into high relief. The immediate future of airborne radar was to be in air-to-surface vessel radar (ASV) and the search for enemy shipping.

Under Robert Watson-Watt, the Research Establishment at Bawdsey Manor had grown so successfully that it was in desperate need of new blood. Its resources were stretched now that the Chain was being built at sites along the coast. Skilled men were needed to supervise construction and to train operational staff. Now that Britain's economy was gearing up to fight a possible war, government and industrial scientists found themselves tied up in war work. Normal recruitment was impossible, as Bawdsey was a well-kept military secret. Few people knew of its existence other than Tizard and the members of CSSAD. It was clear to them that the solution to Bawdsey's shortage of brainpower lay in the academic community. Until now the work had been undertaken largely by electrical engineers, but Tizard understood that academic scientists would bring fresh insights to the new science of range and direction finding. The Royal Society had been asked to take on the task of looking at the universities and classifying the extent and abilities of individuals who could be made available to the war effort. Once suitable candidates had been identified, visits to government research establishments, including Bawdsey, were arranged under strict secrecy for scientists ranging from professors to the youngest, brightest, research workers.

Tizard and his committee worked hard to ensure that radar got the cream of the crop.

In 1938, at the age of forty-one, John Cockcroft was one of the leading lights of the Cavendish Laboratory, Cambridge University's world-renowned physics department. He had joined Cavendish after graduating in mathematics in 1924, a degree he had taken at an older age than average (like so many others) after spending three years as a signaler in the Royal Artillery from 1915 to 1918. At Cavendish he came under the tutelage of Ernest Rutherford, regarded by many as the father of nuclear physics, and in 1932, he and Ernest Walton, under Rutherford's guidance, became the first men to "split the atom" in a controlled manner. He would win the Nobel Prize in 1951 for his work in the use of accelerated particles to study the atomic nucleus. So Cockcroft was ideally placed to be brought into the small circle of Britain's brightest minds who knew about the secret work on what would become the country's radar shield. He recalled:

> My first contact with War Research came through Tizard. We met at lunch in the Athenaeum [Club] and there he talked to me about new and secret devices we were building to help to shoot enemy planes out of the sky. These devices would be troublesome and would require a team of nurses—would we—the Cavendish—undertake to come in and act as nurse-maids, if and when war broke out. He talked also of wanting large powers at short wave lengths and would we think about it.[10]

In early 1938, Tizard promised to arrange for Cockcroft and a handful of other Cavendish scientists to visit Bawdsey to see the work going on, but government security clearance was delayed until Prime Minister Neville Chamberlain had returned from negotiating September's ill-fated Munich Pact, which allowed the German occupation of Sudentenland, with his promise of "peace for our time." Then Cockcroft and his colleague Ralph Fowler drove down to Bawdsey to be initiated into this most secret of war work:

> We had explained to us the elements of pulse technique; how to measure direction and height and how R.D.F. was to be fitted to planes to find ships and other planes and to

> ground stations to shoot guns. We were shown the Bawdsey chain station—with its magnificent 320 foot steel towers for transmitting and its 200 foot wooden towers for receiving. In a cabin in the field we saw the very first G.L. [Gun Laying radar]—being built up by Pollard and Forshaw. We left thrilled with these first visions of a new war science.[11]

To scientists like Cockcroft—especially those who had some wartime experience—the opportunity to get involved at the cutting edge of war science must have been terribly attractive. Nonetheless, CSSAD needed to get the right personnel on board and integrate them with the men who had begun the work and were liable, understandably, to feel a little territorial. That summer Watson-Watt decided to leave Bawdsey for another wartime position, and A. P. Rowe, who had started as CSSAD's secretary, took over as the head of the Bawdsey Research Establishment. Cockcroft and the Cavendish men were still being courted—in typically English style:

> Our next visit was in the spring, the shadow of the Czech disaster over us. I took Dee, Lewis, Ratcliffe, the leaders of the Cavendish. This time we had lectures and all was explained to us at low level by Bowen, Williams, Larnder, Whelpton, Wilkins, the old hands of R.D.F. Bowen showed us his first air borne transmitters for A.I. and A.S.V. We saw also the station on the cliffs which was the first of all coast defence stations. . . . The summer went on, the clouds of war threatening. We arranged for a grand introduction of Physicists to R.D.F.—a party of 80 to be organised to spend a month on chain stations, starting on September 1st—a good choice. A last peacetime visit to Bawdsey culminated in a great cricket match Bawdsey versus Cavendish. I remember Rowe and Larnder playing against us. I scored 30 and we won easily.[12]

As well as Cockcroft and Fowler those who studied with and under Rutherford at Cavendish Laboratory included Otto Hahn, Hans Geiger, Ernest Walton, and Niels Bohr. In 1927, they were joined by a talented and ambitious Australian named Mark Oliphant, who had heard Rutherford speak in New Zealand two years earlier. Oliphant, just twenty-four years old at the time, had been so impressed that he had decided he would do whatever he could to secure a position working in England under Rutherford. At Cavendish, Oliphant found himself working alongside Cockcroft and Walton as they succeeded in splitting the atom. His own work concentrated on the artificial disintegration of the atomic nucleus and designing particle accelerators. He also discovered that heavy hydrogen nuclei could be made to react with each other—the nuclear "fusion" reaction that would be the basis of a hydrogen bomb. Ten years later, the American scientist Edward Teller would draw on Oliphant's work, but in the late thirties, Oliphant, along with so many other nuclear physicists, had no notion that that work would lead to the creation of a new and terrible weapon, saying, in later years, "we had no idea whatever that this would one day be applied to make hydrogen bombs. Our curiosity was just curiosity about the structure of the nucleus of the atom."[13]

In 1937, Oliphant, a dynamic young (still just thirty-six years old) scientist and skilled administrator, landed a new job. He was offered the post of professor of physics at the University of Birmingham in central England where the laboratory was in serious need of a galvanizing influence. Inside two years he had turned the physics department around, attracting a talented group of scientists and laying plans for the building of a cyclotron. Though the work going on in the building would play a crucial part in the development of nuclear weapons, it was not in the field of nuclear physics that Oliphant would make his first mark on the war effort. Instead it was in radar.

As a former Cavendish man, Oliphant and a handful of his top lab members were recruited alongside Cockcroft for that party of eighty scientists that had been organized to spend a month working on-site with the Chain Home stations up and down the eastern coast. The installation allocated to them was at Ventnor on the Isle of Wight, just off the southern coast of England. These visits were of vital importance and were a logical consequence of the lessons that Tizard had learned at Biggin Hill—that those charged with designing and developing new and advanced technological equipment must first understand, preferably at first hand, its operational use. Cockcroft's memoir records that during this time, "Mark Oliphant [was] at Ventnor—on and off—rebuilding the station most of the time."[14]

Eventually, the demands of the laboratory in Birmingham required Oliphant to return, but two of his staff, John Randall and Henry Boot, stayed behind on the Isle of Wight, gaining a deeper understanding of radar in operational use, and insight into the possibilities for future development that would have staggering implications.

The hiatus in the development of airborne interception during 1938 was not the only false move in the technological development of Britain's air defenses during this crucial period. Resources had been diverted down a fundamentally different path for some time. It was a path that would be revealed to be a dead end well before the actual outbreak of war, but there is no doubt that it was a costly waste of time, effort, and money that could have had dire consequences if it had been allowed to go on too long. Today, for example, one can still find in a farmer's field in the rural county of Essex, the remains of an installation which was once part of a larger system covering the entire county. Essex lies on the flight path between Continental Europe and London, to the north of the estuary of the River Thames. The purpose of this derelict construction is at first hard to discern. Astute observers might be able to figure out that it has resemblances to a searchlight station, but in fact this post was responsible, not for searchlights, but for floodlights. This and scores of similar posts were built in the late 1930s with the aim of floodlighting the entire sky above the whole county. The name of the system was Silhouette, and it was hoped that this would be the solution to the problem of defending the United Kingdom against bombers coming by night. The idea was to flood the sky with light from below, allowing high-flying British fighters to pick out the silhouette of the bombers coming in below them. It was a system that perhaps worked better on paper than it would have done in practice, but right up until early 1939 it was expected to be Britain's primary defense against night attack.[15] Fortunately, its capabilities were never tested in battle, because at the beginning of that year, less than nine months before Britain finally went to war, Sir Hugh Dowding, the Commander-in-Chief of Fighter Command, pointed out its fatal flaw. The system, he said, might have stood a chance of working in the previous war, but with the modern generation of fighters it would be to all intents and purposes useless, because the pilot's cockpit sat directly above a broad wing mounted *underneath* the aircraft's fuselage, effectively obscuring the view below. Silhouette was stillborn.

Attention now swung back to the idea of airborne interception. In March 1939, the United Kingdom and France pledged their support to guarantee Polish independence; in late April, the British government began military conscription (albeit on a small scale); and less than a month later Germany and Italy signed the so-called Pact of Steel agreement of collaboration. Much time had been lost getting Britain's night-time air defenses sorted out, and now it seemed there was little left. Fortunately Bowen's airborne team, by now numbering twenty-three, had been making considerable progress. Until that spring they had been testing the equipment in Fairey Battles and Avro Ansons, both aircraft types that were dramatically unsuited to the demands of actual combat. At last, they began to receive Blenheims in which to mount the equipment. Slow and cumbersome it may have been, but the Blenheim was a genuine two-seater fighter, and finally the system began to take on the character of a combat-ready weapon. Here now was a chance to refine the equipment in its proper working environment, and Bowen and his men were looking forward to the opportunity of installing a few prototype models of both AI and ASV radar. What they hadn't expected was that the new urgency meant that an immediate order was placed for thirty sets to be installed in Blenheims, to be supplied to the airborne team in a steady stream from the new factories of the Bristol Aircraft Company. Suddenly what had been for several years a test-bench exercise was turning into the need for mass manufacture.

To make matters even more problematic, this new technology, which had been a state secret closely guarded by a few highly skilled and experienced men, would now have to be taught to the pilots and gunners who would fly these Blenheims into battle. Many of them would be raw recruits, all of them would be getting used to a new aircraft, and now they would have to come to grips with complex electronics and methodologies that must have seemed to have come straight from the pages of science fiction. And who was to teach them these new skills? Bowen's team were working flat out fitting equipment into the new aircraft, a task made even more difficult when the planes started turning up with an elongated nose, which meant all the fittings and cables had to be redesigned and the whole aerial layout changed. But all this work would be in vain unless the crews were properly trained by the time their radar-equipped aircraft became available to them. War appeared to be very close, and the first AI-equipped Blenheim was delivered to No. 25 Squadron at RAF Northolt near London in the middle of August 1939. Bowen's colleague Robert Hanbury-Brown, who had been doing most of the airborne group's flight testing, was deputized to No. 25 Squadron to train the men who would soon use this equipment in combat, largely on the grounds that he was the only man in the country with the necessary experience.

What quickly became apparent to Hanbury-Brown were the terrible limitations of the equipment in actual operation. In radar, the longer the wavelength, the wider the beam. A beam transmitted from an aircraft on a relatively long wavelength fans out so that it would be certain to pick up "ground return," an overwhelmingly powerful echo from a very large object within several miles of the plane itself, namely the ground underneath it, which would completely swamp the tiny echo from any aircraft it might detect. Because of this troublesome issue, the range of this first generation of airborne radar was limited to its height above the ground. So, as Hanbury-Brown's Blenheim flew at 15,000 feet, it was only capable of finding an enemy target within a matching circumference—a range of 15,000 feet, or roughly three miles. Theoretically Chain Home, the defensive radar system, would locate an incoming aircraft and direct the fighter into the vicinity of the raider so it could be picked up by the airborne set. But Chain Home could only fix incoming aircraft to an accuracy of about four miles. With a mile in range to make up, marrying the two systems was a virtual impossibility. A workable night-defense system would clearly require another breakthrough. Fortunately, that September two men were learning the finer points of radar in action in the Chain Home station at Ventnor on the Isle of Wight. Their breakthrough would take radar defense to a new level.

On Saturday September 2, 1939, Britain woke up to the news that Nazi troops had swept into Poland in the wake of a curtain of dive-bombers that had decimated Poland's defenses. The following day, at 11 A.M., BBC Radio fell silent for an announcement by Prime Minister Chamberlain. He informed a somber people that, in observance of their agreements with France and Poland, their country was now at war with Germany. The war that many in Britain had done everything to avoid had finally arrived.

As inevitable as it may have seemed by the end of August there was nonetheless a sense of profound shock as the reality of the situation set in. The war to come was expected to be short and brutal. The twin fears of bombers and gas combined to produce an apocalyptic image of devastation being wrought in the capital. Fresh in the minds of Londoners was the terrifying image of a sky dark with bombers in Alexander Korda's 1936 film *Things to Come*, based on H. G. Wells's apocalyptic vision of the future.

And the air-raid sirens began to wail within an hour of Chamberlain's radio broadcast. It was a false alarm, officially put down to "a single light plane returning from France,"[8] but nevertheless a useful test of how well the air-raid precautions would work. The weekend had already seen a mass exodus from London as carefully laid plans for the evacuation of the capital's children, and others anxious (and able) to escape the threat, were finally activated. Nine of the major arterial routes out of the city were made one-way roads to facilitate the orderly evacuation of up to three million people. Hundreds of thousands of domestic pets were put down in anticipation of the turmoil that would descend on the city after the expected gas attacks. All nonessential railway journeys were now to be curtailed so that the trains could be used to maximum effect. This was the world to which the complacent and elderly British Empire reluctantly committed itself as its government declared war on the most formidable fighting machine the world had yet seen.

Despite the rush to the air-raid shelters in the first hour of the war and the dire forebodings of a quick and bloody conflict, Britain's war got off to a whimper rather than a bang. In fact the period from September 1939 to April 1940 came to be known as "the phony war." There were deaths, of course, particularly at sea where the loss of ships inevitably means heavy loss of life, but the clash of great armies, either on the ground or in the air, simply didn't materialize. In the United Kingdom it was also known as the "bore war" in a tongue-in-cheek reference to Britain's inauspicious Boer Wars in South Africa at the turn of the century. Britain expected the war to come in the air, but Hitler had his hands full elsewhere; and in fact, far from wanting to launch an all-out assault on the United Kingdom, he held out hopes of some kind of a negotiated peace while he consolidated his position in Western Europe.

Phony war this may have been on the homefront, but it was by no means an idle time for the British. In fact it was a vital period for the consolidation of Britain's defenses—the build-up of arms, ships, and aircraft, and also, crucially, the Chain Home system, which would prove to be indispensable when, in the summer of 1940, the Luftwaffe finally arrived in numbers. Radar was a technology that had been experimented with in many different countries, but by the fall of 1939, it had uniquely been developed into a complete system for the air defense of the United Kingdom. The long-range detection of incoming aircraft was no longer just a means of getting people into air-raid shelters, but rather a system for shifting the balance of power between raiding bomber and fighter defenses.

Despite its novel radar defense system, Britain, in 1939, was woefully short of equipment. In the twenty years that followed the end of World War I, Britain's military readiness had suffered from the nature both of the victory and of Germany's defeat.

Hitler, in contrast, had founded the economic recovery of Germany on the development of a new arms industry. Nowhere was the disparity between the two countries' state of readiness more evident than in the air. In September 1939, with the notable exception of the new single-seater fighters the Hurricane and the Spitfire, Britain's air armament had changed little since World War I. For much of the 1930s Britain had been building airplanes largely with a view to using them as bargaining chips in disarmament talks with the Germans—and why invest in new models if you are building planes in order to decommission them? By contrast the Germans, who were starting from scratch, had built a whole air force of modern machines and by the start of the war were turning them out in substantial numbers. Now, as British industrial and military production was stepped up, and with factories and foundries now vulnerable to German bombers, Prime Minister Baldwin's chilling statement from 1932 remained the prevailing view—"The bomber will always get through." Astute thinkers began to argue that, if war was to come, then, as in World War I, Britain would soon find itself in desperate need of industrial resources—from across the Atlantic.

FIVE

The "Special Relationship"

TODAY MUCH IS MADE OF THE "SPECIAL RELATIONSHIP" between the United States and Great Britain. But that special relationship did not exist in the 1930s—certainly not in the form we know it. Of course the two countries had been allies in World War I, but after 1920 the U.S. and Britain had simply stopped cooperating with each other. There was a great deal of mistrust between the two powers—one waxing, the other waning. But in the 1930s, with the growing threat of a new war as Hitler came to power and Germany began to rearm, America's productive power became increasingly attractive as new military demands started stretching British industrial capacity to the limit.

In June 1935 Prime Minister Baldwin moved Philip Cunliffe-Lister, Lord Swinton, from the Colonial Office to the post of Secretary of State for Air. His task was to rearm Britain's Air Force and to equip it with new and modern fighters to defend the country against the growing threat from Germany. A new generation of airplanes was ordered straight from the drawing board—the Spitfires and Hurricanes that would perform so valiantly in the Battle of Britain—to replace the obsolete biplanes that were the legacy of the first war. Britain's rearmament was starting late, but it was in earnest; the largest order that had been placed for airplanes in earlier years was for eighty biplanes; Swinton's first order was for two thousand of the newly designed stressed–skin monoplanes. With the founding of Swinton's Air Defense Research Committee and Tizard's Committee for the Scientific Study of Air Defence, 1935 also saw the beginnings of the development of radar and the new defenses that would serve Britain so well when war finally came, including new bomber designs and advanced power-operated gun turrets and balloon barrages being put into development.

In the spring of 1937, following the abdication crisis and the coronation of King George VI, Baldwin was replaced by the appeasement-minded Neville Chamberlain. Swinton found little appetite for rearmament from the new prime minister. Finance ministers were unwilling to divert peacetime manufacturing resources to war production, and so Swinton persuaded the firms that would be called upon to produce the airplanes if and when war came, to build what came to be known as "shadow factories." When Hitler occupied Austria in 1938, the British cabinet reacted with a further expansion of the Air Force. Now the shadow factories came into their own: production could begin immediately, with fully equipped factories waiting to be put into action. Chamberlain, though, remained committed to his policy of appeasement, stubbornly continuing to believe that a settlement could be reached with Hitler. Swinton was swimming against the prime ministerial tide, and in May 1938, he was sacked by Chamberlain, to be replaced by Sir Kingsley Wood.

Swinton had done much to ensure that Britain had the maximum manufacturing capacity in the run-up to a war, but the problem of maintaining construction after war had broken out remained. Those same factories, indeed all plants engaged in the production of war matériel, would be prime targets for the Luftwaffe bombers. If, when war came, it turned out to be a protracted affair, Britain would surely find itself in critical need of military supplies from elsewhere.

America's factories were three thousand miles from the war zone—immune, it was presumed, from the threat of destruction by

Hitler's bombers. But the purchase of arms was an expensive business, and World War I had left Britain heavily indebted to the United States. Not unreasonably, the Americans were demanding that all military goods be paid for with cash—cash that was in short supply in the UK. Further, it was clear to all that this next war would be a war of new technologies, and that the key to success lay not just in manufacturing capacity but in ensuring that the goods were as technologically advanced as possible. The innovations in the new technology of radar make this abundantly apparent.

Britain faced a twin problem in the 1930s: the need first to provide for a consistent supply of military goods and additionally to ensure that it had all the latest technology coming through in operational form. In this latter respect the British felt they had much to offer arising from their radar research, and much to gain—in particular a top-secret bomb-aiming device developed by the U.S. Navy for its airplanes. Although its details were carefully guarded, its existence and its ability to out-perform all other aiming devices were well-known. The idea began to emerge in the 1930s that it might be possible to exchange information with the Americans on various types of technologies. The trouble was, new military technologies are also top military secrets, and even a potential ally can be unwilling to share them. Especially when that potential ally remains a committed isolationist.

But the two countries had many interests in common, particularly at sea. Across the Pacific, another increasingly belligerent nation, Japan, represented a danger to them both, threatening the British Empire and vital American interests in the Far East. There was much to be said for the two navies cooperating in their attempts to contain the Japanese threat. At the close of 1937, Roosevelt dispatched Captain Royal E. Ingersoll, director of the War Plans Division of the U.S. Navy, to London for meetings with his Admiralty counterpart, Captain T. S. V. Phillips. The task was to divide up the responsibilities of the two navies in the Pacific and to begin the exchange of technical information on communications within the fleet. The success of these meetings in January 1938 prompted the U.S. naval attaché in London to suggest reopening the exchange of technical information between the navies that had been so fruitful during the previous war.

What followed was a series of attempts throughout the year to identify exchanges of information in which both parties felt they were getting a reasonable deal. The difficulties were legion. New military developments tend to be kept closely guarded, and if one side does not know what the other side has, it is very difficult to ask for it, or at least to ascertain what one might be prepared to give in return. Furthermore, each side will regard its own new developments as highly valuable, but will have little information about the value of work being carried out by the other. This was the situation that dogged attempts at a naval technical exchange between Britain and America during 1938.

In May 1938, the Board of the Admiralty in Britain agreed to initiate an informal exchange of information on the basis it described as "quid pro quo." No formal system could be put in place because of the risk of compromising American neutrality, but it was decided that if the Americans requested a particular piece of information it should, in principle, be granted—but only on the basis that something of equal value would be made available in exchange. Toward the end of that month Captain Russell Wilson, the American naval attaché, was given comprehensive details of the Admiralty's two-speed destroyer mine-sweeping devices. In return the British were handed the plans of the U.S. Navy's latest mine-sweeping equipment and access to ships equipped with the experimental gear.

But over the next few months things became increasingly sticky after Captain Wilson went beyond his brief in proposing that the exchange should be extended to cover the area of fleet tactics. This provoked strenuous objections within the Admiralty, and matters deteriorated further in September when the British naval attaché in Washington, who had by this time surveyed the American mine-sweeping equipment, reported that "the gain is entirely to the advantage of the United States," and that it was "in no way an ample *quid pro quo*" for the information that the British had provided. The U.S. Navy wanted information on British boom defenses against submarines, largely because they had none of their own, so they would have to give something else in return. But whatever the British requested, the Americans were unwilling to hand over. With no alternative suggestions being offered, the exchanges were running into rough water. Eventually the Americans agreed to hand over details of their aircraft carrier landing systems, but it had taken almost a year to negotiate.

It was now the spring of 1939, and the prospects for any further successful naval exchange were worse than ever. The two leading figures at the Admiralty had both been replaced by Chamberlain appointees. Duff Cooper, the First Lord of the Admiralty, had resigned in protest over his government's appeasement of Hitler at Munich, while Lord Chatfield, the First Sea Lord, had retired late in 1938. Their replacements were far less inclined toward technical exchange with the Americans. But the Royal Navy was not the only arm of the British military interested in American technology. The RAF had been after something for quite a while too.

In 1932, the U.S. Army Air Corps had taken delivery of their first Norden bombsights, the Mk XV. The inventor C. L. Norden had been perfecting bombsights for the U.S. Navy's airplanes since 1920, and in the Mk XV, which came to be known as the M-1 in Army service, it had reached an unprecedented standard of accuracy. The Norden bombsight was a quantum leap ahead of any of its competitors. It was essentially an analog computer combined with a sophisticated optical sight, which could take account of variables like altitude, airspeed, and wind velocity to calculate the correct drop point. Other bombsights could do that, but what made the Norden so special was that it was gyroscopically stabilized and coupled with an automatic pilot, so that once the sight was locked on to its target the plane would be guided precisely to the drop point. It was the U.S. military's superweapon, and the press made much of the fact that America had it.

The forthcoming war was anticipated in Britain, as elsewhere, as a bombing war, so for the RAF this was a very desirable technology. First attempts to acquire it had been made as early as spring 1938, but the RAF had been told the Norden was not for sale. Later that year George Pirie, the British air attaché in Washington, approached the U.S. Army with an offer to exchange bombsight technology. Again the British were rebuffed, this time with the Army pointing out that they could not help because the rights to the bombsight rested with the Navy.

The RAF had nothing to match the capabilities of the Norden, and were determined to continue with efforts to get the Navy to part with it. Those efforts, however, had to wait after a spy scare erupted in Washington over the leaking of classified material to foreign attachés. After a hiatus of six months the offer of exchange was renewed—this time with the British prepared to throw in some additional technology of their own. One vital piece of equipment the British had developed that the Americans had not was the ability to mount hydraulically powered gun turrets on their bombers. To the British this was a crucial design feature of the modern bomber, which would be expected to be able to defend itself once out of the range of escorting fighters. This new offer, which came in the early months of 1939, might have had some chance of success in Washington, but for a curious incident that took place a few weeks earlier.

On January 23, a military plane plunged into a parking lot in Los Angeles and burst into flames. The pilot had bailed out at three hundred feet, too low for his parachute to deploy. His flight engineer did not eject and he, too, was killed when the plane hit the ground. The aircraft was a prototype of the latest twin-engine Douglas DB-7 light attack bomber straight from the company's drawing board and, unusually, there was one other man on board. Flying in the fuselage toward the back of the plane, he survived the crash and managed to escape before the wreckage caught fire. He was, however, badly injured, suffering a broken leg, severe back injuries, and a slight concussion. The lucky man was described by Douglas as a mechanic called Smithin. Journalists, however, soon discovered that Smithin was in fact a French Air Force officer, Captain Paul Chemidlin, attached to the French purchasing commission. It quickly emerged that he had been given permission to fly in the new plane by President Roosevelt himself, against Army advice. The incident proved almost as uncomfortable for Roosevelt as it had for Chemidlin. The isolationists were convinced that Roosevelt was trying to drag the United States toward war with the Fascists, and used the event as proof that in pursuit of that aim he was quite prepared to jeopardize national security by selling precious military secrets. This was not the ideal climate for a technical exchange involving the

Norden bombsight, widely thought of as America's most closely guarded military innovation.

The RAF appetite was whetted still further by a demonstration of the capabilities of the Norden bombsight to which the British air attaché George Pirie was invited on April 13, 1939, at Fort Benning, Georgia. Four B-17s were scheduled to fly across the target area at an altitude of 12,000 feet. Their task was to drop bombs on the target of a "battleship" outlined on the ground. Pirie's account of the day makes impressive reading. "The first B17 was due to drop its bombs at 1.27 P.M. About 1.26 P.M. everyone started to look and listen for it. Nothing was seen or heard. At 1.27 while everyone was still searching the sky, six three hundred pound bombs suddenly burst at split second intervals on the deck of the battleship, and it was at least 30 seconds later before someone spotted the B17 at 12,000ft."[1] It was no fluke. The other three bombers all hit the target too. Pirie was mightily impressed and wrote to London urging that everything possible should be done to lay hands on the Norden bombsight.

But the climate for exchange was becoming ever more difficult. The Chemidlin incident had set Roosevelt back and strengthened the hand of the isolationists. Meanwhile in Europe on March 15 the last vestiges of any hope that the Munich agreement might have curtailed Hitler's expansionism vanished when his troops marched into what remained of Czechoslovakia. War between Germany and the allied forces of Britain and France was beginning to look unavoidable. The prospects for an exchange deal bringing the Norden bombsight to Britain were now looking equally doomed. If war did indeed break out and British bombers equipped with the Norden bombsight were to be used to attack German targets it was inevitable that one would eventually be shot down. If that were to happen and the bombsight were to fall into German hands, the United States would have lost its technological advantage without ever having been involved in any conflict.

By late May it was clear that attempts to obtain the Norden bombsight for use by the RAF had finally run aground, and George Pirie reported as much to London, recommending that any future attempts to obtain the device would have to be made at a diplomatic level. In fact it appears that an unexpected offer had already been made using diplomatic channels. On May 1, 1939, Joseph Kennedy, the U.S. Ambassador, had gone to the Air Ministry in London to meet with the new Secretary of State for Air, Kingsley Wood. According to Kennedy's telegrams back to Washington, Wood had floated the idea of an exchange of radar technologies—a remarkable gesture considering that Britain had been making great progress in this field. However serious or official the offer, in fact it got nowhere at all. Harry Woodring, the then U.S. War Secretary, rejected the idea out of hand, partly on the basis that he doubted the British scientists had anything better than their American counterparts, and partly because his hands were tied by public promises made by Roosevelt following the Chemidlin incident.

Acutely aware of their bombers' failings, the British decided that, with German troops now massing on the Polish border, the request had to be made at the very top. On August 25, 1939, Prime Minister Chamberlain wrote directly to President Roosevelt:

> I make this urgent personal request to you because Great Britain today faces the possibility of entering on a tremendous struggle confronted as she is with a challenge to her fundamental values and ideals. Moreover, I believe they are values and ideals which our two countries share in common. . . . Should the war break out, my advisors tell me that we would obtain a greater immediate increase in our effective power if we had the Norden bombsight at our disposal than by any other means we can foresee. Air power is, of course, a relatively new weapon which is so far untried on a large scale; there is the danger of unrestricted air attack which we for our part would never initiate. I am however most anxious to do all in my power to lessen the practical difficulties which may arise in operations against legitimate military targets, and feel that in air bombardment accuracy and humanity really go together. For this reason again I am certain that you would render the greatest service if you could enable us to make use of the magnificent apparatus which your Services have developed.[2]

Roosevelt rejected Chamberlain's request, citing existing neutrality laws and security legislation in the United States that meant that the only way the bombsight could be provided was if it was made available to all other foreign powers, a situation that was of course unacceptable on both sides of the Atlantic. The following day German forces crossed the border into Poland to trigger a second world war.

With the outbreak of war in September, Winston Churchill was summoned in from the cold by Chamberlain to take up his old position as First Lord of the Admiralty. In America, Britain had a new ambassador, one whose diplomatic expertise would prove vital in laying the groundwork for the negotiation of scientific exchange. "We had at this time in Washington a singularly gifted and influential ambassador. I had known Philip Kerr, who had now succeeded as Marquess of Lothian, from the old days of Lloyd George in 1919 and before," recalled Churchill, describing a man with whom he had "differed much and often from Versailles to Munich and later."[3] Lothian had been appointed ambassador to Washington in August 1939, just a few days before Churchill reentered the cabinet. Lothian's position was vital, a key channel of communication to the upper echelons of power in the U.S. He was in many respects the ideal man for the job.

Lothian was, in fact, not a professional diplomat, but his connections in America were at the highest level. He was secretary to the Rhodes Trust, an endowment that had taken many of America's brightest postgraduate students on scholarships to Oxford. He was on friendly terms with the influential journalists William Allen White and Walter Lippmann, as well as with Felix Frankfurter, whom Roosevelt had appointed to the Supreme Court. His cousin, Mark Kerr, was an old friend of Roosevelt's, and he had met the president personally on a number of occasions. Perhaps Lothian could persuade the Americans to part with their most precious secret in return for the information they now sought on Britain's sonar antisubmarine defense system, ASDIC.

In fact, that information had already been offered—through different, and highly informal, channels. When Churchill returned to the Admiralty in September 1939, he had more or less immediately opened up a personal exchange of cables with President Roosevelt. It was the beginning of a correspondence that would serve the allies well throughout the war. But there was clearly some potential awkwardness following from the fact that a cabinet minister had a more direct and personal line of communication with the U.S. head of state than the prime minister. Consideration was given to stopping the communication, but in the end it was decided that there was more to gain by keeping it open, provided those in more official channels knew what was going on. It was essential to avoid the diplomatic embarrassment of this correspondence contradicting the position being taken at the ambassadorial level. So even though the communication remained strictly personal, it found its way across the Atlantic via official channels at both embassies. Kennedy and Lothian, the two ambassadors, were to be kept informed at all times.

One of its early fruits had been the tacit agreement under which in late October the United States unilaterally established a three-hundred-mile-wide neutrality zone around the coast of the Americas, a zone into which the ships of belligerent nations were forbidden entry. On the face of it this applied as much to the Royal Navy as it did to the German, whereas in fact Churchill had privately agreed to it as part of a plan for dividing up the policing of the oceans between the navies of Britain and the United States. But Churchill had gone further, and had, without apparently consulting his colleagues, offered to pass Britain's ASDIC secrets to the U.S. Navy. Nothing had specifically been asked for in return.

Talks were quickly initiated, with Captain Alan G. Kirk, the U.S. naval attaché, sitting down with the Admiralty in London in early November. Admiral John Godfrey, the director of naval intelligence at the Admiralty, was shocked by Kirk's request and made it clear that the Admiralty were only prepared to hand over ASDIC secrets as part of a reciprocal exchange of anti-submarine technologies. By November 27, London's position had hardened. Once again it was the Norden bombsight that was requested as the American side of the bargain. And once again this was unacceptable to the U.S. Navy, despite the fact that Britain's Air Ministry now claimed to have a device that would destroy the Norden bombsight if a bomber was shot down. In early December Lothian was assigned to take this new offer of ASDIC and details of the new destruction device for the bombsight directly to the president. Meanwhile Britain's War Cabinet had been considering a new formal American offer of a like-for-like swap of anti-submarine technology. So unappetizing did they find it that they decided not to send a formal rejection, but rather to hope that there would soon be a positive response to Lothian's offer.

On December 13, Lothian and Roosevelt met in the White House. Roosevelt listened attentively, clearly intrigued by the idea of a new device that would offer protection for the Norden bombsight in combat conditions. But he remained unwilling to allow the British access to the bombsight, which had now become something of a political hot potato. As Lothian left the meeting his proposal was still on the table. No definitive response from the White House ever came.

Britain, then, waited on the Americans, while Washington waited on a British response. Formal attempts to initiate an exchange of technological secrets at both military and diplomatic levels had reached a stalemate, and minds were now elsewhere—Britain preparing for hostilities in continental Europe, while the U.S. was considering the implications of a major violation of the new neutrality zone along the Atlantic seaboard of the Americas. On the coast of Uruguay, British warships had cornered the German pocket battleship *Graf Spee* in the mouth of the River Plate, forcing its captain to scuttle the ship. The combatants in a European war were slugging it out on America's doorstep. If exchange were to happen it was beginning to look as if it would have to come by way of a different route.

By 1939, in his fifty-fifth year, Sir Henry Tizard had a distinguished career behind him as an airman, academic, and skilled administrator. His stock was high with the men of influence in all three of those communities. But he was also, and primarily, a scientist. As Britain went to war, then, Tizard was ideally placed to know what the military (and in particular the RAF) needed, and also what science could do to help. And in Professor Archibald Vivian Hill he had a colleague of enormous ability. Hill was a physiologist who had been awarded a Nobel Prize in 1922 for work on the production of heat in muscles. His stature as a scientist was international. But he, too, was experienced in military matters. During World War I he had been one of Britain's leading lights in the development of anti-aircraft gunnery.

On the CSSAD he had been Tizard's staunchest ally, particularly in the run-in with Lindemann.

Two years earlier, in 1937, Tizard had proposed that a scientific attaché should be assigned to work alongside the military attachés at the British Embassy in Washington. His job would be to investigate American scientific and technical developments and the possibilities of collaboration. This proposal from Tizard's (purely advisory) committee was rejected on the grounds that the successes at Bawdsey would mean that the Americans stood to gain much more by such a two-way agreement. Now, after the outbreak of war, Tizard had revived this proposal, and Hill had agreed to take on the task. By the end of 1939, attempts at organizing the exchange of information at a military and diplomatic level had ground slowly to a halt. Tizard's proposal would tackle the matter on a totally different plane—direct communication and cooperation between scientists in two different countries, setting aside chauvinistic interests in pursuit of developing new technologies that the military research facilities were unable to produce on their own. Tizard's aim in sending Hill to America was simple but immensely sophisticated. He feared that Britain would soon find itself cut off from continental Europe and without allies, and that its capacity to manufacture weapons in the quantity required would be quickly exhausted. Tizard saw that Britain needed access to the technical resources and productivity of America. It was a process of procurement that Swinton had already started for conventional weapons, but Tizard knew that the key lay with the newer technologies in the vital field of electronics—particularly those in air defense, where Britain almost certainly had a head start over the U.S.

By early March 1940, Hill was ready to take up an attachment to the embassy in Washington as supernumerary Air Attaché. His purpose was to investigate American attitudes toward the war and to build bridges at the highest levels among their scientific, political, and military elites. While he was at it, he was to try to find out what progress, if any, the Americans were making on radar or any of the other secret devices the British war effort had spawned.

Hill's attachment was not speedily organized; even in wartime the British civil service could manage a considerable degree of wrangling over who would pick up the bill for Hill's new post and to whom he would report in Washington. But Hill complicated matters further when he decided in early 1940 to stand for a seat in the House of Commons. When he won the election on February 25, 1940, it became constitutionally difficult for the Foreign Office to pay him as he was now a member of the legislature, a different branch of government. Compromise, though, is one of the things the British are rather good at, and within a few days it had been arranged that his salary would be paid by the Royal Society, Britain's premier scientific institute, and he would not take up his seat in the House of Commons until he returned from America in three months. Meantime Hill had put the delay to good use. In January he had begun writing a series of letters to leading British scientists asking for recommendations as to who he should try to see when he arrived in United States. Hill wanted names and addresses, and if they had personal knowledge of the contacts they were recommending, he wanted to know what position they took on the war. From these and other sources he assembled an eleven-page list of key contacts. It encompassed scientists, scientific administrators, and executives at major radio technology manufacturing companies.

With everything finally settled, Hill set sail for America on March 9, 1940. Though events in Europe were now unfolding fast, one thing that could not be hurried was the transatlantic crossing. Virtually no paperwork survives from Hill's trip to America, and he himself admitted, in later correspondence with Tizard, that his memory for details was vague. Of one thing, though, he had a clear memory. It's a curious insight into how a Nobel prizewinner alleviates the boredom of a ten-day sea voyage: "I was at sea on 18 March, the day of the real Equinox. I wanted to know where we were—about nine days out—and timed the sunset. This gave me the longitude since the sun set that day at the same time independent of latitude."[4]

Hill arrived in Washington at the start of the last week in March, and before the month was out Lord Lothian had arranged for him to meet with Felix Frankfurter who, he wrote later, was "very helpful with the President."[5] Frankfurter told Hill that he thought Roosevelt would be receptive to the idea of an open and frank technical exchange, and that he would do everything he could to help secure the president's approval.

Another name high on Hill's list was that of Vannevar Bush, the senior U.S. scientific administrator—in many ways Tizard's counterpart in America at this time. Bush, the chairman of the National Advisory Committee for Aeronautics and president of the Carnegie Foundation, was sympathetic to the idea of scientific exchange, and he helped smooth the path as Hill began a hectic round of visits to military and industrial research facilities. Hill's inquiries turned out to be more interesting than even he and Tizard could have imagined, not so much for what he was able to find out as for the things to which he was denied access. Hill was hoping to visit radio research facilities, and after the positive tone of his meetings with Frankfurter and Bush he was optimistic about the prospects. However he soon discovered that even these friends in high places were no guarantee of free access to the information he sought. A request to visit RCA's Research Center in Camden, New Jersey, was turned down flat, and on his tours of the General Electric and Bell Telephone laboratories he was accompanied at all times by a naval officer whose task was to ensure that he had only limited access. An attempt to set up a meeting with Dr. Frank Jewett, the head of Bell Laboratories, to discuss the possibility of providing the British with information on American aircraft detection equipment was turned down by Jewett on the grounds that the Navy had specifically banned him from talking about such equipment. Against all British expectations it seemed that the Americans might be developing their own radar systems, and indeed that they might have something to offer that the British did not have.

During the ten weeks or so of his trip Hill was in regular contact with Tizard, writing privately to him back in London. Tizard noted some of the correspondence, which reveal the development of a new understanding that attempts to "trade" secrets were simply getting in the way of any meaningful exchange, and that, ultimately, Britain's interests might best be served by laying its cards openly on the table. At one point Hill had a relatively informal meeting with Dr. Frank Aydelotte, the president of Swarthmore College. Tizard's archives record Hill's report of a telling exchange between them: "[Aydelotte] said to Hill, 'What are your people doing with radio waves reflected from aircraft for directing anti-aircraft gunfire?' and added 'Don't tell me if you think you ought not to.' AV's opinion was 'We are fooling ourselves if we put any trust in secrecy. It is far more important, if possible, to keep ahead of the enemy. We can get very considerable help in the United States if we are prepared to discard what is by now really a fetish.' Later on he said 'Our impudent assumption of superiority, and failure to appreciate the easy terms on which closer American collaboration could be secured may help us to lose the war.'"[6]

Hill was convinced that Britain might be damaging itself by protecting "secrets" the Americans already had. He was sure there was significant work on radar applications going on behind the scenes in civilian laboratories in the U.S. Tizard recorded that on April 23, Hill wrote privately to say that "the Americans were onto a large number of applications of RDF, but neither (Pirie) nor himself got any impression that the service people know about them." He also records that "a remark by [Vannevar] Bush left him in no doubt that the services are developing, or beginning to develop, a number of RDF applications."[7] Hill's researches were suggesting that Britain might have more to gain by surrendering its secrets than merely accessing the productive power of the United States. Hill felt he had seen enough of the American scientific community to know that there was tremendous goodwill toward the British cause, and that there was huge potential for research, development, and production on that side of the Atlantic if the British were prepared to hand over their secrets and, crucially, if it was done in the right way.

Hill sent this encouraging news back to Tizard; in addition, he made an important recommendation—that no attempt should be made to trade secret for secret with the Americans. Subject to appropriate security arrangements, he advocated that there should be a complete disclosure of British defense secrets up to that date. He was confident that the Americans would react to this in a generous and wholehearted way.[8]

But there was still the tricky issue of the political position. Would such a gift of British secrets be acceptable to the still neutral United States? Would the president's private wish to be supportive of Britain be compromised by such a mission? Would it be seen as potentially providing powerful ammunition to the isolationists? Perhaps

Lothian, the British ambassador, would be able to sound out the president personally. Late in April, Hill wrote to Lothian outlining his position and advising that "It might be of major importance to offer a very frank interchange of technical knowledge and experience as a basis for the indirect co-operation which promises to be so valuable and so unlikely to arouse political objections."[9]

Lothian was more than happy to put his weight behind the plan, and on April 23, he cabled his support to the Foreign Office in London, along with a summary of Hill's findings. He told London that he thought there was an additional benefit to the plan at the diplomatic level in that it might bring "the services of the two countries into closer liaison and sympathy [on] war preparation."[10]

Nonetheless, Lothian's telegram betrays the fact that the idea of bartering secrets had not entirely disappeared. He wrote: "Our present bargaining position is strong owing to [our] earlier start and service use. If we wait the USA will probably discover the essential for themselves. I suggest that we offer full information on RDF [radar] and its developments in exchange for similar information on their systems and for complete facilities to obtain the latest types of instruments and equipment to our specifications and requirements from firms now working on United States Government contracts." He went on to say that he "would try to get the release of the bomb sight as part of the exchange though as a method of negotiation it might be wiser not to [do] this at first."[11]

Though it clings on to the idea of exchanging radar secrets for the Norden bombsight, Lothian's cable is significant because it identifies the emergence of two quite distinct aims—access to America's secrets, and access to its factories. Back in London the Foreign Office got behind their ambassador's proposal. The day after receiving Lothian's cable the Foreign Office's American department notified the Air Ministry: "We are all in favour of such an exchange of information especially as the Ambassador seems to think there is some chance of getting the Norden bombsight as part of the exchange."[12]

Two days later the matter was brought up in the War Cabinet. The issue was left unresolved, however. Tizard continued lobbying for an open technical exchange, calling in support for the idea from senior scientists in just about every building in Whitehall. In the

Admiralty, however, there was considerable apprehension about the idea of handing over secrets to their American counterparts. Following Tizard's promptings, a meeting was called on May 3 at the Air Ministry to which representatives of all three services were invited. The aim was to open discussions on precisely *how* technical and scientific cooperation with the United States might be set in motion. Admiral James Somerville was less than keen, expressing a view widely held in the Navy that "anything told to the American Navy went straight to Germany." This grave allegation of lax American security was based on little more than rumors from naval attachés in Washington. But for now, yet another meeting had ended inconclusively. All that had been achieved was a request to Tizard to produce a report summarizing the pros and cons.

Though personally convinced that the pros far outweighed the cons, he nevertheless made a laudable attempt to summarize both arguments. But he was at pains to point out in the report that the basic theory of radar was well known throughout the world, so that it was quite possible that the Germans already had a system of their own, and in any event an aircraft carrying air to surface vessel radar had recently been shot down so it was quite possible that the Germans were already in possession of some of Britain's very latest equipment. The report concluded, though, clearly in favor of sending a technical mission, and made the key suggestion that serving officers from all three branches of the military should be included as well as civilian scientists. But Tizard's proposal was about to be overtaken by events.

SIX

Prime Minister Churchill

In May 1940 the Phony War came to an abrupt end when German soldiers swept into the Netherlands, Belgium, and Luxembourg. The battle which had begun on the other side of Europe in Poland was now on Britain's doorstep. Dawn broke on May 10 to the start of a brutal assault by fast, heavily-armored tank divisions supported by dive-bombers and paratroopers. Heavy bombers laid waste the great cities of the Low Countries. After just four days a third of Rotterdam lay in ruins, a thousand of its people dead, 85,000 homeless.

The Germans had gained confidence and strength and were aggressively expanding the fight into western Europe. Almost exactly one month earlier Germany had launched its invasion of Norway and Denmark. Norway had been a much sought after prize. The Germans had two principal designs on Norway: They were anxious to secure the supply of iron ore from Norwegian mines at Gällivare, and they regarded control of the country's coastline as a vital extension of the front on which they could challenge the British by sea and in the air. The long, jagged coastline, pierced as it was by deep and well-protected fjords, offered shelter to warships and submarines. They would provide the ideal base from which to attack Allied convoys and, when the time came, the British fleet itself.

None of this, of course, was lost on the Allies. Britain and France had long contemplated the benefits of securing control of Norway for themselves. The difficulty lay in Norway's neutrality. A plan was hatched to enter Norway under the pretext of sending a force to support the beleaguered Finns in their war against the Russians. This scheme was thwarted, however, by the refusal of the Norwegians and Swedes to allow an Allied Expeditionary Force to enter their territory. Cautious voices also counseled against the plan for fear of provoking a military collision with Russia. The idea languished before being rendered moot when the Finns surrendered to the Russians on March 12. A chance, albeit a dangerous one, had been lost.

A new tactic was now required, and its architect was Winston Churchill, now Britain's First Lord of the Admiralty. He proposed laying mines to prevent the free movement of German shipping in Norway's protected coastal waters and to force ships into international waters where they could be engaged. German retaliation would give the Allies the opportunity to send troops to occupy the ports along the Norwegian coast. After some wavering on the part of the French, the plan was scheduled for April 8. In later years it would become known that German intelligence was able at this time to intercept British signals and was aware of the Allied designs. Once again Hitler would strike the first blow.

In the days leading up to April 8, British reconnaissance had observed heavy movements of German shipping in Norwegian waters. The fleet now put to sea in expectation of forcing a naval engagement on the high seas reminiscent of World War I. The German Navy, however, had no such plans. On the morning of April 8, the Allies triumphantly announced the successful mining of Norwegian waters. The Norwegian Foreign Office began preparing a protest against what they saw as a violation of their neutrality. But worse was to come. In the early hours of April 9, news reached Britain of German troop landings at Trondheim, Bergen, Stavanger, and Oslo. The mines were useless; the Germans were already there. Within a matter of hours they were in control of southern Norway.

Over the next month the situation in Norway would turn into a complete disaster for the Allies. The campaign would expose weaknesses in military planning as well as a political leadership totally unsuited to the demands of a fast-moving war. A British attempt to retake Trondheim, regarded by the Norwegians as the key to preventing the German armies from moving northward and gaining control of the rest of the country, was bungled, and the British were about to learn the bitter lesson of the importance of tactical air support. The Germans, thanks to their occupation of the mainland, had it. The British, hundreds of miles from their nearest air bases, did not. Within a week conditions had deteriorated so badly that plans were being prepared for the almost total evacuation of Allied troops from Norway. Germany had struck first and the Allied riposte had been repulsed. Allied vacillation, together with the speed and adroitness of German tactics, had turned Norway into total defeat. The blame was laid squarely at the door of Chamberlain and his War Cabinet. A campaign was under way to replace him.

British politics, though parliamentary, is primarily a two-party system, and in the strained economic climate of the 1920s and 1930s the battle lines of British politics were clearly drawn. On one side were the Conservatives, the party of old money and the middle classes. On the other was the Labour Party, representing the workers and committed to moving the country toward a socialist system of government. The differences had been increasingly emphasized during the interwar period as much of the world polarized into fascism or communism. Yet by the spring of 1940 such was the peril in which Britain stood that all sides had begun to believe that the time had arrived for a united national government. As the debacle in Norway unfolded, so the momentum built for a wartime restructuring of government. Chamberlain's stock, however, was at its lowest ebb, and Clement Attlee and the other leaders of the Labour Party had made it plain that while they supported the idea of a National Government and would agree to take part, they would, under no circumstances, be prepared to do so with Chamberlain as prime minister. Lord Halifax,

Chamberlain's Foreign Secretary, was the obvious choice, and very much favored within the Conservative Party. But he could not command the support of the Labour Party, and on May 9, he told Chamberlain that if he accepted the post, he would be the prime minister "in name only." Winston Churchill was everyone's choice to be given charge of defense, and Halifax felt that this situation would effectively put Churchill in charge of running the war rather than himself.

As this momentous day wore on, Churchill, despite by his own admission bearing much of the responsibility for events in Norway, began to emerge as the candidate behind whom everyone could unite. Both major parties and the minority Liberals were happy to enter a government of national unity with Churchill as its head, believing that, whatever his past mistakes, his personal vigor and appetite for the fight were precisely what was needed to replace the indecision that characterized Chamberlain and his cabinet. Just as vitally, these qualities were recognized by the population at large. By the evening of May 9, all seemed to be settled and Churchill stood ready to take charge of his country in its hour of need. He went to bed that night planning the composition of the War Cabinet he would attempt to put together the following day. Yet again, Hitler would intervene. There would be twenty-four more hours of turmoil before Churchill assumed office.

France's Maginot Line, the massive fortification along its border with Germany, had been built in the 1930s to ward off a direct assault by Germany. In rebuilding Germany's armed forces, Hitler had created an army that was modern and highly mechanised. He and his generals had been quick to see its possibilities and had developed a new sort of warfare, the blitzkrieg, which they had honed to near perfection in the invasion of Poland the previous year. The key to blitzkrieg was the fast and agile Panzer divisions, equipped with heavily armored tanks working with close air support from fighters and dive bombers. The Maginot Line, in contrast, was built to deal with the sort of war of attrition that had characterized the 1914–1918 conflict. Hitler's plan was simply to go around it through the thick woods of the Ardennes. For weeks German troops had been massing along the borders with Belgium and the Netherlands. The Low

Countries were all that stood in the way of a high-speed sweep into northern France.

The threat was clear, but the Allies had been handicapped in their response. The powerful French Army, more than two million strong and augmented by the British Expeditionary Force of over 300,000 men, under the command of the veteran General Maurice Gamelin, was desperate to draw up a defensive line opposite the Germans. Caught between the two mighty European powers of Germany and France, Belgium and the Netherlands had sought to maintain an uneasy neutrality after the armistice of 1918. Despite the obvious threat, their governments now continued to maintain strict neutrality in the hope that this would dissuade the Germans from any aggressive move. To have taken the other course and opened their borders to the French Army would have been to abandon their neutrality and invite German retaliation. The best the French could achieve was to establish a line along their own border with Belgium, effectively making that country a buffer zone between the two armies.

Watching events unfold in Britain, Hitler must have felt that this political instability was an opportunity too good to miss. The bulk of the French Army he knew to be badly positioned. More than half its 103 divisions were stationed in or behind the Maginot Line, and its six armored divisions, which alone were equipped to meet the Panzer divisions on equal terms, were dispersed along the whole length of the front. The British, with their threat of a substantial air force and the world's largest and most powerful navy, looked, for the moment, entirely rudderless. Now was the moment to strike.

Just as Churchill lay pondering the composition of his cabinet, German troops were beginning to pour across the Belgian and Dutch frontiers following swift air attacks against airfields and lines of communication. Both countries had sought security in a policy of firm neutrality. Now they were being invaded along a front stretching for 150 miles. On May 10, their nightmare had come to pass. The reaction of the Allies was swift, but in the end ill judged and badly executed. The Belgian border was suddenly opened and the Allied armies rushed forward to try to halt the Germans in their tracks.

To Chamberlain, for now still prime minister, the matter was simple. He set his views out in a letter to Lord Beaverbrook "As I expected Hitler has seized the occasion of our diversions to strike the great blow and we cannot consider changes in the government while we are in the throes of battle." For several hours during that dark May morning the political situation remained fluid, with Chamberlain trying to gather support for the maintenance of the status quo while Churchill pressed his case that these events made early changes in the government all the more vital. Things looked as though they could go either way until the country's senior socialist leaders weighed in decisively: Labour would serve in a government of national unity, but not under Chamberlain.

Churchill had succeeded in ousting from the head of the War Cabinet a man whose name was inextricably linked to the discredited policy of appeasement. But Churchill recognized that casting Chamberlain aside would risk alienating the majority party in the House of Commons and would deny his new government a serious source of influence and expertise. When Churchill appeared before King George VI to accept the awesome responsibility of leading a War Cabinet, he was flanked by Chamberlain and Halifax, as well as Clement Attlee and Arthur Greenwood, Labour's two most senior Members of Parliament. That night Churchill retired to bed at about three in the morning. "I was conscious," he said, "of a profound sense of relief. At last I had the authority to give directions over the whole scene. I felt as if I were walking with Destiny, and that all my past life had been but a preparation for this hour and for this trial."[1]

The tenth day of May 1940 must count among the most momentous days in world history. The German invasion of the Low Countries was a disaster for the Allies. In little more than a fortnight the British Expeditionary Force and large numbers of French soldiers had been thoroughly routed by the Germans, forced into retreat, and trapped on the beaches of Dunkirk. In what would become known as "The Miracle of Dunkirk" well over 300,000 British and French soldiers were evacuated. First calculations seemed to suggest that the ships available to the Admiralty were sufficient to rescue only about 45,000 men, less than one-sixth of those taking refuge on the beaches. Beginning on the morning of May 27, ordinary seafaring Britons began to assemble at ports along the southern coast of England opposite northwestern France and the beaches of

Dunkirk. Over the next few days they would build to a vast armada of some 860 vessels of all shapes and sizes that would ferry the soldiers to safety. The soldiers were forced to abandon all their equipment, and the defense of France itself was fatally compromised, yet the new prime minister somehow managed to turn this reverse into a source of new strength for the British. The loss of equipment at Dunkirk was a major setback. The loss of 300,000 men would have been a disaster. British morale had been badly dented by the debacle of Norway; to lose an entire army of men would have dealt it a fatal blow. These 300,000 men were an invaluable asset which would return to haunt the Germans.

The combination of Dunkirk and Churchill's arrival in 10 Downing Street, however, seemed to provide an extraordinary opportunity for the people of Britain to rally round the charismatic figure of their new prime minister and, at last, to do something about the events unfolding in Europe. Dunkirk seemed to forge a new spirit of resistance among ordinary Britons. It was a spirit that had its apotheosis in Churchill himself:

> There is no doubt that had I at this juncture faltered at all in the leading of the nation I should have been hurled out of office. I was sure that every minister was ready to be killed, and have all his family and possessions destroyed, rather than give in. In this they represented the House of Commons and almost all the people. It fell to me in the coming days and months to express their sentiments on suitable occasions. This I was able to do because they were mine also. There was a white glow, overpowering, sublime, which ran through our Island from end to end.[2]

Sir Henry Tizard may not have felt a white glow on hearing of Churchill's arrival at Number 10. Tizard was well into his fifth year in a position of powerful if quiet influence over the technological development of air warfare. Working with a hand-picked team who shared his vision, Tizard had been able to support the fledgling radar scientists and, crucially, to focus their work in the directions that would add most to the coming war effort. Such was their success during this period that hardly a blind alley was followed. And, apart from the short period when Lindemann had been added like a cuckoo in the nest, the Tizard Committee had been able to follow these goals with precious little interference. He had succeeded in promoting and seeing through to fruition the Chain Home system, which would provide Britain's first line of defense. The system now extended the length of the southern and eastern coastlines of Britain, almost ready for the stern test it would face in just a couple of months' time. But Tizard also had the foresight to realize the implications of this system and to set in motion the embryonic development of airborne radar, even at a time when the huge physical structures needed to produce a radar trace were far too big even to contemplate their installation in aircraft where weight and space were at a premium. Lord Swinton's committee, the Air Defence Research Committee, had executive powers in its role as a subcommittee of the Committee for Imperial Defence, and Swinton had succeeded in pushing millions of pounds through government for this work. But when Churchill took the reins of power on May 10, Tizard knew things were about to change.

Churchill was a naval man at heart. He had been First Lord of the Admiralty in World War I, and he had been brought back to the post by Chamberlain on the declaration of war on September 3. But his appointment in 1935 to Swinton's Air Defence Research Committee had given him five years of experience at the sharp end of air policy. In this, as ever, Lindemann had been a constant source of advice, and after the ignominy of exclusion from Tizard's Committee he carried a burning sense of injustice that his voice was not being heard. Now, as his political comrade assumed power, that voice was going to be heard at the most influential levels. So significant was this relationship that Lindemann would remain one of Churchill's closest advisers throughout the war. J. G. Crowther has observed: "He was an essential part of the Churchillian system. If Churchill was the leader Lindemann had to be accepted. Lindemann could not be removed without removing Churchill."[3]

After May 10 events quickly began to undermine Tizard's position. One of Churchill's first acts as prime minister, intended to accelerate the supply of Britain's war planes, was to strip that responsibility away from the Air Ministry. He created a new ministry, the Ministry of Aircraft Production, bringing in the charismatic newspaper baron Lord Beaverbrook as its head. The new ministry would have responsibility for all aircraft acquisition and development. This alone left Tizard somewhat high and dry in his position as scientific adviser to the chief of the Air Staff; but what made matters worse was that Beaverbrook and the new Secretary of State for Air, Sir Archibald Sinclair, were close political allies of Churchill. It was obvious where they would look for scientific advice, and that was to Lindemann rather than Tizard. By early June this situation had crystallized when Tizard discovered from Robert Watson-Watt that a crucial meeting had been called at the Air Ministry to discuss future research and development work. Tizard's name was conspicuous by its absence from the list of invitees. Air Ministry officials told Tizard this was because the meeting was of little importance, but Tizard attended anyway. He was surprised to find that among those at the table was the new Secretary of State, Sinclair, with, as his principal scientific adviser, Frederick Lindemann. On June 19, after a second meeting had taken place at the Air Ministry with Lindemann present and Tizard excluded, Sir Henry was aware that his position was becoming untenable.

Two days later Lindemann's power was conclusively demonstrated at a meeting that took place in the Cabinet Room at Downing Street. Lindemann brought in a bright young man in intelligence research at the Air Ministry. His name was R. V. Jones and he claimed to have identified the existence of a German precision bombing system. The system, code-named *Knickebein*, or "crooked leg" for the shape of the antennas, used a focused radio beam to guide bombers toward their target. A second, intersecting beam would tell them when they had arrived at the precise moment for bomb release.

Jones was well aware that German prisoners, largely pilots whose planes had been shot down, could be a useful source of information, and so he went to great lengths to ensure a supply of information from Britain's intelligence-collecting agencies. In early 1940 news came through of an overheard conversation between prisoners in which a system called *X-Gerät* (or X-Apparatus) was mentioned, in terms which implied that it was some sort of blind bombing system. Jones was further encouraged in this hunch by the salvaging of a document from the wreck of a shot-down German aircraft in March 1940 that spoke of a radio beacon "operating on 315°."[4] Jones surmised that the only purpose of specifying a bearing for a radio beacon would be if it were transmitting some kind of directional beam. On May 23, 1940, he summarized his convictions in a report to be sent to the prime minister, among others, in which he set out his belief that the Germans had a beam system that could guide their bombers along a path that would be no more than half a mile wide over London. It was a potentially devastating aid to bombing accuracy, and it was this memo that led to Jones being summoned to explain his theories to the prime minister and his cabinet.*

If Jones was correct, the implications were huge. No longer would the bombers be restricted to clear, moonlit nights to finds their way to targets in Britain. With such a system they would be able to navigate blind, flying through heavy overcast conditions, hidden from the defending fighters, searchlights, and anti-aircraft batteries. Now Britain's cities would be vulnerable in all weathers. Winston Churchill recounts the story in his memoirs:

> Lindemann told me also that there was a way of bending the beam if we acted at once, but that I must see some of the scientists, particularly the Deputy Director Of Intelligence Research at the Air Ministry, Dr. R. V. Jones, a former pupil of his at Oxford. Accordingly, with anxious mind, I convened on June 21st a special meeting in the Cabinet Room, at which about fifteen persons were present, including Sir Henry Tizard and various Air Force commanders. A few minutes later, a youngish man—who, as I afterwards learned, thought his sudden summons to the Cabinet Room must be a practical joke—hurried in and took his seat at the bottom of the table. According to plan I invited him to open the discussion.[5]

*Jones's secretary inadvertently typed half a meter instead of half a mile, a degree of accuracy which would have been frightening indeed. Fortunately, none of the recipients seems to have remarked upon this error.

Jones told the prime minister that sources on the continent had reported that the Germans were developing a new system for night bombing. At first it had been thought that it involved German agents placing beacons in the target cities. But then, a few weeks earlier, two or three oddly shaped squat towers had been photographed near the coastline facing Britain. Their shape and position could not be explained by any known technology. Then an enemy bomber had been shot down equipped with apparatus more elaborate than Lorenz, the Luftwaffe's night-landing system. Reports of the interrogation of a German pilot seemed to confirm that the towers had something to do with a blind-bombing system.

Churchill recalled: "When Dr. Jones had finished there was a general air of incredulity. One high authority asked why the Germans should use a beam, assuming that such a thing was possible, when they had at their disposal all the ordinary facilities of navigation. Above twenty thousand feet the stars were nearly always visible. All our own pilots were laboriously trained in navigation, and it was thought they found their way about and to their targets very well. Others round the table appeared concerned."[6] Whether Sir Henry Tizard was that "one high authority" or not is impossible to say; but he was skeptical of Jones's research and very much opposed to too many resources being diverted into it. Churchill, though, had heard enough to convince him. He immediately ordered that the existence of the beam should be assumed and counter-measures should be given the highest priority.

Later that same day, June 21, Tizard tendered his resignation. Clearly his advice no longer carried any weight. On June 26, Tizard wrote to all members of the Committee for Scientific Survey of Air Defense, informing them that he was resigning as scientific adviser to the Chief of the Air Staff and as chairman of the committee, and explaining that he felt that his ability to advise had been compromised along with that of his committee: "With the change of Government matters have become worse in this respect, not better. There are too many advisors and too much conflicting advice; responsibilities are not clear cut, and in consequence action is slow. . . . I have therefore resigned my post as Scientific Advisor, and as Chairman of the Committee, I have also told the Secretary of State that I do not think that the Committee can continue to be of much help unless there is a change in the organisation."[7]

Within just a few weeks Jones would be proved conclusively right when the German began *Knickebein*-guided raids and the British, through Jones's insight, were able to some considerable degree to thwart them. Tizard had been wrong, and now it was Tizard who was out in the cold.

SEVEN

The Common Cause

WITH FRANCE FALTERING, BRITAIN STOOD ALONE, but with a new spirit and, thanks to Churchill, a vigorous determination to prevail. The road, he knew, would be long and hard. There would be difficult decisions to take and setbacks aplenty, but Churchill was fortified with the idea that he was fighting to preserve a way of life, the way of liberal democracy, a way epitomized in the Anglophone democracies on both sides of the Atlantic. Born to an American mother, he felt a strong bond with the United States and was convinced they would eventually join the battle against fascism. From the moment he came to power this firm conviction would do much to condition his decisions, and much to support him through Britain's darkest hours. For the next eighteen months he would do everything he could to try to draw on the economic, industrial, and military strength of America—sure that, despite the powerful isolationist lobby, the United States must see that if Britain fell, and its naval control of the Atlantic fell with it, the land and liberty of America would be left vulnerable to a now-mighty Fascist war machine intent on world domination.

On May 15, just five days into office, Churchill had made a plea to Roosevelt for "the loan of 40 or 50 of your older destroyers to bridge the gap between what we have now and the large new construction we put in hand at the beginning of the war."[1] At the end of the previous war the Royal Navy had boasted more than four hundred destroyers: now, facing the prospect of imminent invasion and the urgent need to retain control of the English Channel, it had the services of only sixty-eight seaworthy destroyers. The idea was to fit these aging four-stack, flushdeck destroyers—veterans of World War I—with ASDIC, the latest submarine sonar detection device, and bring them quickly into home waters to deter the impending invasion. Roosevelt had been sympathetic, but he had not found a way to help without provoking resistance in Congress.

The relationship between these two men and these two countries would define the rest of the war, but for now Britain was alone and exposed—and there was worse to come. After Dunkirk the Germans had pushed on toward Paris. The valiant resistance of the French armies became more and more disorganized, with communications breaking down and supply lines failing as the roads became swamped with refugees fleeing before the German advance.

During these desperate few weeks France was appealing for help wherever it could. Churchill flew five times to France to meet Prime Minister Reynaud and French leaders, who were pleading for British air support. Several times he put his own life at risk, his light plane surrounded by a squadron of Hurricane fighters. This was a momentous decision facing a prime minister still little more than four weeks in the job. If air support could really turn the tide then that would best serve the Allied (and Britain's) cause. Churchill's inclination was to agree with Reynaud's view that this was the fulcrum of the war, and that unqualified British air support could tip the balance. He convened a special cabinet meeting, inviting Sir Hugh Dowding, the commander-in-chief of Fighter Command, to join them. Churchill was dismayed at Dowding's contribution. Dowding spread before him a graph showing the current disastrous rate of attrition of the Hurricane fighters that had already been lost over France. The line of the graph sloped steeply downward. Dowding told the prime minister: "If the present rate of wastage continues another fortnight we shall not have a single Hurricane left in France, or in this country." So complete was Churchill's power at this early point that he was unaccustomed to not getting his own way. Indeed even the Air Minister, Sir Archibald Sinclair, and the Chief of the Air Staff, Sir Cyril Newall, were inclined to give in to Churchill's instincts. But now here stood the man with day-to-day responsibility for the fighters and the men who flew them saying that acceding to the French request would leave Britain defenseless. Dowding had taken a vital stand, and Churchill gave in. Dowding advised Churchill that he must hold a minimum of twenty-five fighter squadrons in reserve for the defense of Britain, and that is what he did. Air support would be forthcoming, but by no means to the degree the French wanted.

As the month of June marched on, and the tragedy of France unfolded, fascism seemed to be running rampant, sweeping away the liberal democracies in its path. On June 10, Italy invaded southern France and declared war on Britain. Mussolini's opportunistic entry into the war just as the Germans were closing on Paris was pretty much the last straw for Roosevelt. On this day his private conviction that America would have to join the war effort became dramatically public. He had agreed to give a commencement address at the University of Virginia, where his son Franklin Jr. was studying law. Roosevelt decided to adopt in his speech Reynaud's description of the Italian action as a "stab in the back." But this was too much for his cautious State Department officials, who removed the phrase from the speech. Roosevelt was not happy though, and on his way to Charlottesville, after much thought, he put it back in. Joseph Lash reports Roosevelt's wife, Eleanor, as saying that, having done so, he felt "a sense of relief and satisfaction."[2] It was the phrase by which this speech would be remembered, though its true significance lies in the passage when he returns to the issue of "material resources":

> On this tenth day of June, 1940, the hand that held the dagger has struck it into the back of its neighbor.
>
> On this tenth day of June, 1940, in this University founded by the first great American teacher of democracy, we send

> forth our prayers and our hopes to those beyond the seas who are maintaining with magnificent valor their battle for freedom.
>
> In our American unity, we will pursue two obvious and simultaneous courses: we will extend to the opponents of force the material resources of this nation; and, at the same time, we will harness and speed up the use of these resources in order that we ourselves in the Americas may have equipment and training equal to the task of any emergency and every defense.[3]

Roosevelt's speech must at last have given Churchill some hope for American support in the war. Mussolini's action threatened to change the military map of Europe at a stroke. The Mediterranean Sea was a vital supply line for Britain. Ships bringing goods, or even military reinforcements, from its empire in the East could cut some 8,000 or 9,000 miles off the journey by using the Suez Canal before traversing the Mediterranean from east to west, exiting into the Atlantic through the Straits of Gibraltar. Protecting this route had been a key goal of the British Navy for more than a century, using bases in Suez, Malta, and Gibraltar. With the tacit support of the navy of close ally France, this control of the Mediterranean had been largely unchallenged. Now, however, with France seemingly on the verge of formal capitulation, this whole imperial supply line was coming under grave threat. Italy had a powerful, modern navy, with its home bases right in the heart of the Mediterranean. The vista that this opened up for Churchill was truly appalling. If France capitulated, Tunisia, Morocco, and Syria would all go with it. With Italy now on Hitler's side there was a clear threat of a German push through the Balkans toward the vital oilfields of the Middle East, bypassing Turkey through the French possessions in the Levant and Palestine. Britain would be left alone defending the Mediterranean, Suez, Malta, and Cyprus, as well as the Near East and Egypt. The troop ships and merchant vessels sailing from the empire in the east would soon be exposed to great danger of attack.

But worse yet was possibly in store. Little more than a week had passed since the almost miraculous rescue of 300,000 British troops from the disaster of Dunkirk. The army had been saved, it was true, but the tools of their trade had been abandoned in northern France. Once France fell the victorious German troops would be ideally placed to extend their push across the English Channel, where the defending forces of their only remaining opponent were cruelly weakened. The German Navy would be in full control of the continental Channel ports, while the Luftwaffe could supply air support from the captured airfields of northern France. On the night of June 10, 1940, a German invasion looked more than likely. A mere twenty-odd miles of sea held the German war machine at bay.

Roosevelt's remarks had been bold and principled, but politically dangerous. In one speech he had risked alienating both the isolationists and the strong Italian American lobby with that dramatic phrase "we will extend to the opponents of force the material resources of this nation." The beleaguered Churchill was acutely aware of the potential significance of Roosevelt's words, and before going to bed that night he dictated a heartfelt cable to be sent the following morning: "We all listened to you last night and were fortified by the grand scope of your declaration. Your statement that the material aid of the United States will be given to the Allies in their struggle is a strong encouragement in a dark but not unhopeful hour. Everything must be done to keep France in the fight and to prevent any idea of the fall of Paris, should it occur, becoming the occasion of any kind of parley." But the immediate material aid that Churchill had in mind was those World War I destroyers he had requested from Roosevelt. He took the opportunity to renew his plea: "Not a day should be lost. I send you my heartfelt thanks and those of my colleagues for all you are doing and seeking to do for what we may now, indeed, call the Common Cause."[4]

Whether Roosevelt would have called it a "Common Cause" is a moot point. He had stubbornly resisted the release of the destroyers in the month since Churchill had come to power. To have sanctioned their transfer would have given powerful ammunition to the isolationist cause. Yet in wanting to help the British he had, at the end of May, approved the sale to Britain of twenty motor torpedo boats that were being built for delivery to the U.S. Navy in July. Aware of the political consequences of even this small measure, he had taken care to clear it first with Navy Chief Admiral Harold Stark.

But Roosevelt had reckoned without Senator David Walsh of Massachusetts, the chair of the Senate Naval Affairs Committee and a staunch isolationist. Walsh questioned the legality of diverting vital resources away from the U.S. Navy. The matter was raised with the attorney general, who concluded that it was indeed in violation of the Neutrality Acts. Roosevelt was obliged to backpedal and cancel the deal. It was onto this less than fertile ground that Churchill's renewed plea for destroyers fell. On June 13, Roosevelt would sign the Navy expansion bill (with the amendment proposed by Senator Walsh) authorizing $1.3 billion to be spent on additional construction of warships. By the end of June the Senate had amended the Navy expansion bill so as to forbid the disposal of any Army or Navy matériel unless the chief of staff or the chief of naval operations should certify them as "not essential to the defense of the United States."[5] Meanwhile, in response to Churchill's pleas in his telegrams, a vast stock of weapons desperately needed by the British to replace those left behind at Dunkirk were about to be shipped to Britain on the SS *Eastern Prince*—500,000 rifles, 129 million rounds of ammunition, and 80,000 machine guns along with many artillery pieces. This formidable array of weaponry had been assembled from U.S. government stores and certified as surplus to U.S. requirements. The Neutrality Laws had been subverted by first "selling" the arms to a steel company and then reselling them to the British government. As June came to a close Roosevelt signed another Navy Bill providing $550 million for the construction of forty-five more ships and other projects.

The diplomatic cables crossing the Atlantic from both Britain and France now became desperate, yet still failed to elicit aid. On June 10 French Prime Minister Reynaud cabled both Churchill and President Roosevelt to describe the new threat to France and to ask again for help. Churchill took the opportunity to cable Roosevelt the following day in support of the French and to ask for an American squadron to be sent to Ireland for fear that the Germans might attack on that flank. That night he flew to France. On his return he again cabled Roosevelt. The French position was deteriorating and now was the time for America to come to its aid. His greatest fear, and one that he knew would resonate with Roosevelt, was that the French fleet would fall into German hands, wresting control of the Atlantic away from the Royal Navy.

On June 12, 1940, Churchill cabled Roosevelt warning that the aged Marshal Pétain was advocating peace with Germany, while Reynaud and a young General Charles de Gaulle were prepared to fight on. He entreated Roosevelt to strengthen Reynaud's position and encourage the resistance. The following day, Roosevelt cabled his reply to Reynaud. Ambassador Joseph Kennedy delivered a copy of it to Churchill in London.

Roosevelt's statement that "It is most important to remember that the French and British Fleets continue [in] mastery of the Atlantic and other oceans; also to remember that vital materials from the outside world are necessary to maintain all armies" was by no means the support the French were looking for, and for France all this came too late—German tanks rolled into Paris the following day. But it must have provided encouragement for Churchill to think that Britain, too, might be able to count on "vital materials from the outside world." It was a clear indication of Roosevelt's view of the naval situation, that it was crucial for the Allied fleets to retain control of the Atlantic, leaving the U.S. fleet free to concentrate on its western defenses against the growing threat of the Japanese in the Pacific.

Churchill eventually had to make the difficult decision that the remainder of the British Expeditionary Force, 150,000 men under General Alan Brooke fighting alongside the French in northern France, should be withdrawn. Brooke advised that the situation was hopeless and that it was best to withdraw the men and their equipment to fight another day. The evacuation of Brooke's troops through the Normandy ports, along with some 20,000 Canadians, had already begun. Some 156,000 men were safely brought back to Britain, although in one terrible incident the 20,000-ton liner, the *Lancastria*, with 5,000 men on board, was bombed and set on fire just as it was about to leave port. Over 3,000 men lost their lives in this disaster, news of which Churchill suppressed for several weeks for fear of the damage it would do to Britain's otherwise surprisingly strong morale.

Now suddenly it seemed the French, or at least army officers and senior members of government, had lost the will to fight on and were looking to sue for peace in an attempt to avoid being overrun by the Germans. On June 17, the news came that Marshal Pétain had asked for an armistice and ordered all French troops to stop fighting. The alternative, to fight on and possibly take the fleet overseas, was looking increasingly unlikely. Unlikely too that the fleet would be scuttled, as putting the fleet at their disposal would be a key element of German armistice terms. France had now effectively collapsed, and Britain stood exposed, facing the dread prospect of a combined French and German Navy. Once again Churchill was called upon to lift the spirits of the British people in the face of a terrible reverse. He stood to address the House of Commons on June 18. He had come to inform them of the events of the previous two days, and that French resistance was at an end. He did so in somber terms, but ended his speech with words that gave strength to the British for the fight to come:

> What General Weygand called the Battle of France is over. I expect that the Battle of Britain is about to begin. Upon this battle depends the survival of Christian civilisation. Upon it depends our own British life, and the long continuity of our institutions and our Empire. The whole fury and might of the enemy must very soon be turned on us. Hitler knows that he will have to break us in this Island or lose the war. If we can stand up to him, all Europe may be free and the life of the world may move forward into broad, sunlit uplands. But if we fail, then the whole world, including the United States, including all that we have known and cared for, will sink into the abyss of a new Dark Age, made more sinister, and perhaps more protracted, by the lights of perverted science. Let us therefore brace ourselves to our duties, and so bear ourselves that, if the British Empire and its Commonwealth last for a thousand years, men will still say "This was their finest hour."[6]

On June 24, General Huntziger, leader of the French delegation sent to finalize the terms of France's surrender, signed the armistice with Germany. In a move designed to rub French noses in the ignominy of defeat, the same railroad carriage that had been the site of the 1918 Armistice at Compiègne had been specially retrieved from its pride of place in a museum celebrating the Allied victory. An armistice ending the two-week-old war with Italy was concluded two days later.

Over the next two weeks the British made several entreaties to Pétain's government to hand over the French fleet, or at least to put it beyond the reach of the Germans. But despite assurances, nothing was done. Roosevelt meanwhile, fearful of an even greater disaster if Britain should fall, was keen to seek assurances from Churchill that the British fleet would meet no such fate. Churchill's request for American destroyers still hung in the air. Now, on top of domestic opposition, Roosevelt had the fresh fear that destroyers sent to Britain in this way would soon find themselves under German command. Churchill's resolve to fight on was strong, but the prospect of a German invasion was very real, and the loss of matériel at Dunkirk had, for the time being, weakened Britain's ability to resist. Plans were being made to take the government into exile if that were to happen, probably in Canada, and in that event the fleet would be sailed to new bases across the Atlantic.

But Churchill was disinclined to give Roosevelt the public assurance he sought. The Dunkirk spirit still sustained the British, even after the fall of France, but Churchill knew that it rested on delicate foundations. To publicly state that the fleet would be sailed to North America in the event of an invasion would send out the wrong signals and risk undermining that precious morale. On June 28, he set out his position in a cable to Lord Lothian, his ambassador in Washington. Churchill wanted to be sure that the president understood the subtleties of his position.

> Prime Minister to Lord Lothian (Washington), 28 June 1940
>
> No doubt I shall make some broadcast presently, but I don't think words count for much now. Too much attention should not be paid to eddies of United States opinion. Only force of events can govern them. Up till April they were so sure the Allies would win that they did not think help necessary. Now they are so sure we shall lose that they do not think

> it possible. I feel good confidence we can repel invasion and keep alive in the air. Anyhow, we are going to try. Never cease to impress on President and others that if this country were successfully invaded and largely occupied after heavy fighting some Quisling Government would be formed to make peace on the basis of our becoming a German Protectorate. In this case the British Fleet would be the solid contribution with which this Peace Government would buy terms. Feeling in England against United States would be similar to French bitterness against us now. We have really not had any help worth speaking of from the United States so far. We know President is our best friend, but it is no use trying to dance attendance upon Republican and Democratic Conventions. What really matters is whether Hitler is master of Britain in three months or not. I think not. But this is a matter which cannot be argued beforehand. Your mood should be bland and phlegmatic. No one is down-hearted here.[7]

Five days later, on July 3, the British set out to neutralize the French Navy. A number of different ways were offered by which the ships could be put beyond German control, ranging from sailing them across the Atlantic to French colonial ports in the West Indies to scuttling them in the harbors when they lay. None was accepted, and at 6 P.M. "with profound regret" the British Navy opened fire on those who had so recently been their comrades in arms. At Mers-el-Kebir near Oran in Algeria, Admiral Somerville destroyed much of the French fleet as it lay in harbor. The *Bretagne* was sunk and two more battleships badly damaged. The *Strasbourg* and five destroyers managed to break out of the harbor and set sail for Toulon. Two more French battleships, nine destroyers, and many smaller ships at anchor in the British ports of Plymouth and Portsmouth were occupied by British troops, mostly without resistance. Further skirmishes now followed British attacks in North African harbors where other small flotillas of French ships lay in port. In Alexandria the French commander, Admiral Godefroy, allowed his ships to be demobilized. The battleship *Lorraine*, three heavy cruisers, one light cruiser, three destroyers, and a submarine were taken out of active service. In less than a week the danger of a substantial French Navy continuing to operate under the flag of Vichy France had been defused through a combination of military force and negotiation. It was a decisive pre-emptive strike, and one that Churchill felt sent out the most resounding signal of Britain's intentions.

The fear of the French Navy falling into the hands of the Germans had been laid to rest, and Churchill had made it privately clear to Roosevelt that no such fate would be allowed to befall the British fleet. So what now for the prospect of practical support from the United States? The "vital materials" of Roosevelt's cable to Reynaud on the eve of the fall of Paris must have occasioned something of a wry smile for Churchill. Up to now such materials had not proved as easy to deliver as President Roosevelt might have hoped.

Roosevelt was a shrewd and pragmatic politician, and right up to the end of the 1930s he was perfectly sure that the strong strain of isolationism among the American public, faithfully reflected on Capitol Hill, meant his chances of pushing repeal of the Neutrality Acts through Congress were just about nil. He would have to wait until events in Europe began slowly to shift the direction of public opinion and force the hand of Congress. In the spring of 1940 that time had come. But now as those events began to move at startling speed there was a grave and growing danger that their pace would outstrip the ability of America to respond. For some time now Roosevelt, though neutral in public, had been to a considerable degree interventionist in private. From the moment Hitler's troops entered the Low Countries in early May, so Roosevelt had begun to act.

Roosevelt was approaching the end of his second term in office, and presidential elections loomed in November 1940. Many had been urging him to stand for an unprecedented third term. He had, for some time now, been convinced that in one way or another, the war was coming to America. And, if he was right, then America would need a new, and unique, government of national purpose, and a strong and charismatic president who could carry the people with him in just the way Churchill was in Britain. Sometime in late May or early June, between the invasion of the Low Countries and the fall of France, judging that public opinion was on the move and that events in Europe were coming to a climax, Roosevelt decided to allow his name to be put forward for the Democratic nomination. At the same time he swiftly began to remold his administration, shaping it for the new circumstances of an approaching war. In Britain Churchill had come to power as part of the creation of a national government in which all parties were represented. Britain, though, was at war, whereas most Americans still thought they would manage to stay out of it. But Roosevelt knew he had both to bring experience into his team and to court the conservative vote. So in June 1940 he made the remarkable and quite unexpected announcement that he was adding two new senior Republican figures into his cabinet. Colonel Frank Knox, once a close associate of Theodore Roosevelt, was appointed Secretary of the Navy to replace Charles Edison, regarded by Roosevelt as ineffective. Even more extraordinary was the appointment of Henry L. Stimson to the post of Secretary of War. Stimson had served in the same position under William Howard Taft, and had then been Secretary of State in Herbert Hoover's administration. He was something of a Republican grandee, and his appointment to this influential post alongside Knox would do much to signal a sense of national unity in the way Roosevelt proposed to react to the events unfolding in Europe.

Roosevelt was not alone in sensing that public indifference to the war in Europe was disappearing fast. In the middle of June, Lord Lothian wrote Nancy Astor, a U.S.-born member of Britain's House of Commons, "the change has been staggering in the last fortnight, it would take very little to carry them in now—any kind of challenge by Hitler or Mussolini to their own vital interests would do it."[8]

Back in Europe, German forces began their occupation of the Channel Islands on June 30. This was the only British territory they would conquer, but in that blighted summer it must have seemed like the thin end of the wedge. British forces were soon stretched to the limit. In the Mediterranean the Royal Navy's Group H Task Force found itself engaged by Italy's strong air force and the powerful navy: with two battleships, and four more on the way, along with a powerful force of cruisers and destroyers and the largest submarine force in the world, Italy's complement of fighting ships was more than a hundred strong. As many as 300,000 Italian soldiers stationed in Libya were expected imminently to begin a push along the coast into

Egypt, with the object of capturing the fertile Nile Delta and eventually the Suez Canal itself. British troops in North Africa were outnumbered by five or six to one, yet successfully mounted a series of small skirmishes aimed at disrupting Italian plans. In English coastal waters and the western approaches the Luftwaffe's bombers and the German U-boats, now operating freely in the Bay of Biscay, were taking a heavy toll on merchant shipping and on the destroyers assigned to protect them.

In Britain itself there was the threat of imminent invasion. With the American rifles still on their way onboard the *Eastern Prince*, the defense of the homeland at this stage rested largely with the Local Defence Volunteer force, a motley crew of the elderly and infirm known colloquially as "Dad's Army," and armed with pickaxes and sporting guns. On July 10 the bombers arrived. Seventy German planes droned across the Channel to raid dock targets in South Wales. Churchill's Battle of Britain had begun.

EIGHT

A Simple Stroke of Genius

While Continental Europe had been collapsing under the Nazi onslaught, British RDF scientists had been taking enormous strides, both in terms of practical systems, like Chain Home, and startling advances in the lab. Edward Bowen had been at work on the problem of airborne radar from the very beginning. But there were two major problems. The first was that the equipment used in Chain Home was huge—remember John Cockcroft had been awestruck by his first sight of Bawdsey's "magnificent 320 foot steel towers for transmitting and its 200 ft wooden towers for receiving." The transmitter itself weighed several tons, while the receiver was a huge rack of equipment, a spider's web of interconnecting valves, control knobs, and indicators that required a highly trained operator. The second problem was intrinsic to the way radar itself worked.

The origins of radar defense in Britain went back to the beginning of 1935 and the early days of the Tizard committee. The difficulty was, as calculations made by Watson-Watt's assistant Arnold Wilkins showed, that the strength of the signal reflected from an aircraft would be tiny by comparison with that which would be reflected back from the ionosphere. The first system constructed at Orfordness clearly reflected its origins in the National Physical Laboratory. Operating on a wavelength of fifty meters, it unsurprisingly proved at first to be better at detecting the ionosphere than passing aircraft. But it didn't take the team long to recognize that shortening the wavelength would produce far better results. The bright young Welshman Eddie Bowen was a key member of that early team, and he explained in his book *Radar Days* the push for ever-shorter wavelengths:

> Since the first success at detecting aircraft with radio pulses at Orfordness in 1935, the trend in operating wavelength was steadily downward. Starting at 50 metres, the wavelength was changed first to 26, and then to the band from 10 to 13 metres. Experimental systems were also built at Orfordness and Bawdsey at wavelengths down to 4 metres, followed in the summer of 1937 by widespread use of 1.5 metres for airborne and other applications. With the components available in Britain at that time, both the transmitted power and the receiver sensitivity of a radar system fell away in an alarming fashion at wavelengths below 1.5 metres and for a while this became the lower limit. However, we all dreamed of shorter wavelengths and narrow beams.[1]

Eddie Bowen dreamed with good reason of "shorter wavelengths and narrow beams," because that's exactly what would be needed to put radar into planes, and by the time the Orfordness pioneers had moved to Bawdsey Manor, Bowen was head of a small group trying to crack the problem of airborne radar.

The second of those two major problems about putting radar into planes is that radio waves are just that—waves. Imagine you and a friend are holding a sheet stretched taut between you. Now if one of you begins to shake the sheet up and down a series of waves will be transmitted along the sheet to the other end. The speed at which you shake is called the "frequency" (the number of waves per unit of time) and the wavelength (the distance from crest to crest) is inversely proportional to this frequency. The faster you shake, the shorter the waves. Slow the frequency down and the waves get both longer and wider from top to bottom.

In radar then, the longer the wavelength, the wider the beam, and here is the second problem with putting radar into airplanes. A beam transmitted from an aircraft on a relatively long wavelength would pick up ground return, obscuring the tiny echo from any aircraft it might detect. Narrow the beam, however, and increase the power, and airborne radar would begin to look like a genuine potential addition to the air defenses. That required shortening the wavelength to something that could be measured not in meters but in centimeters, also known as "microwaves." Also a tiny power source would be needed capable of generating these tiny wavelengths with enough power to guarantee reflections from objects hundreds of miles away. The trouble was that, as 1938 rolled into 1939, and Hitler's aggressive intentions became more and more apparent, none of this seemed remotely possible.

Bawdsey at this time was working tirelessly on the Chain Home system, a frantic race to get it ready before the expected war began. Bowen found little institutional enthusiasm for trying to conquer this seemingly intractable problem of generating microwaves at sufficient power: "there was practically no interest within Bawdsey itself and anyone talking about centimeter waves was thought of as some kind of crank."[2]

But radar developments at Bawdsey were being closely followed not just by the Air Ministry, but by the Admiralty too, and Bowen found he was able to discuss this problem a number of times during the spring and summer of 1939 with Sir Charles Wright, the director of scientific research for the Admiralty. Wright, however, wore not just a naval hat. He was also the chairman of the Committee for the Co-ordination of Valve Development (CVD), which had responsibility in the field of electronic development for all three services, Army, Navy, and the Royal Air Force. Wright was quick to see the potential of a narrow beam system for airborne radar and asked Bowen to come up with a prescription: "I did a back-of-an-envelope calculation of the best wavelength for the purpose. . . . The only way of improving the system would be to project a narrow beam forward to get rid of the ground returns. I estimated that a beam width of 10 degrees was required to achieve this. Given an aperture of 30 inches—the maximum available in the nose of a fighter—this called for an operating wavelength of 10 centimetres."[3]

Bowen's calculations were not only prophetic, they were also the stimulus behind a contract placed by the CVD with Mark Oliphant's physics laboratory at Birmingham University, along with two other labs at Bristol and Oxford. In fact, so all-encompassing would this work become that the Birmingham lab's name was changed to the Admiralty Research Laboratory. It was in this laboratory that John Randall and Henry Boot, returning from the Chain Home research station at Ventnor, would, within just a few months, make the breakthrough that made 10 centimeter "microwave" radar possible and with it change the face of war.

After their lengthy stay at Ventnor, in the fall of 1939 Randall and Boot had returned to their laboratory at Birmingham University in Britain's midlands full of the enthusiasm of bright young men exposed to a new challenge. Unfortunately, because they had been away from base longer than most of their fellow scientists, Oliphant had already assigned to others the key work of trying to find a powerful source of 10 centimeter waves for transmitting radar. Instead they were given the job of developing a receiver using a readily available technology called the Barkhausen–Kurz valve. They had some success, but soon realized that they had no good way of testing their receivers without the ability to transmit.

The long-standing search for something capable of producing microwaves at very high power had so far produced a number of disappointing alternatives. Among them was the magnetron, a special form of diode developed some twenty years earlier by Albert Hull of the General Electric Research Laboratory in Schenectady, New York. In recent years the magnetron had been overtaken by a new device with seemingly greater potential. It was called the klystron amplifier, and it had been invented by brothers Russell and Sigurd

Varian at Stanford University. As 1939 drew to a close Oliphant's team was focusing on trying to coax more power from the klystron, which as yet was woefully insufficient to power microwave radar. The magnetron was basically a vacuum tube inside a magnetic field, which caused electrons passing through the tube to follow a curved path. The trouble was it could not produce much more than about 30 or 40 watts at centimetric wavelengths—nowhere near the power necessary. The Klystron looked inherently more efficient, but there was simply no way of getting enough power through it. Randall and Boot decided to look at the possibility of combining the two in some way.

During the summer of 1939, the thirty-four-year-old John Randall had taken his wife and young son on vacation to the Welsh seaside resort of Aberystwyth. On a visit to a secondhand bookstore he had come across an English translation of Heinrich Hertz's seminal work *Electrical Waves.* This book includes an account of the famous spark gap experiment in which Hertz succeeded in generating high-frequency radio oscillations across a gap in a simple loop of wire. What would happen, Randall asked himself, if the loop were simply extended into three dimensions? Hertz's work showed that the wavelength of the electromagnetic power from such a resonator would be exactly 7.94 times the diameter of the loop. Back in Birmingham, Randall discussed the matter with Boot and the two men decided to assume that this figure would remain the same even if the loop were extended to become three-dimensional, like a piece of drainpipe. Knowing that the target the laboratory had been set was to develop a transmitter which would produce microwaves at the 10 centimeter wavelength, they rounded Hertz's results up to 8, and concluded by a simple piece of arithmetic that the diameter of the cavity (the three-dimensional loop) would have to be 1.2 centimeters.

The next breakthrough was a simple stroke of genius. Why not combine a number of these cavities around a central cathode to multiply the effect? Six small holes in a short cylinder of metal, exactly like the cylinder of a Colt revolver—indeed some slightly fanciful accounts say that's exactly what Randall and Boot used as a jig. The way it works is very much like a whistle, where air oscillates as it passes over a gap. But in the cavity magnetron it's electrical energy, not air, that oscillates.

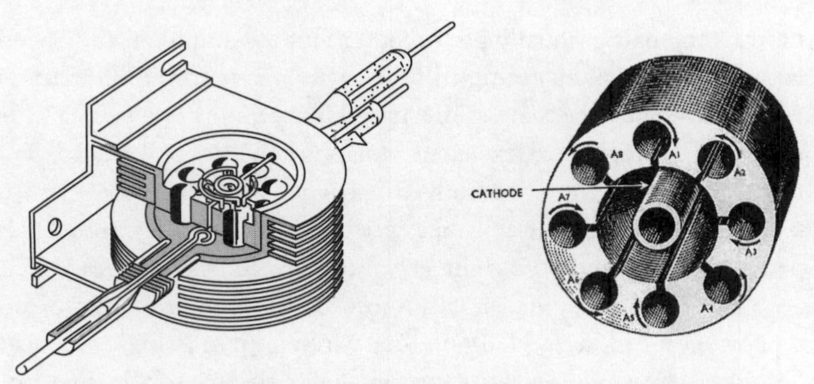

The cavity magnetron is a simple-looking device whose operation is rather complex. Holes, or cavities, are located in a ring that acts as the positively charged anode; a rod through the center of the anode acts as the negatively charged cathode. A magnet surrounds the anode so that as an electrical current is applied to the magnetron, the magnetic field causes the electrons to circulate, rather than shoot off in straight lines from the magnetron. As they move in a circle the electrons pass over the cavities and create an electrical frequency based on the size of the holes. This frequency is conducted away from the magnetron in microwaves. (*National Research Council*)

In simple terms, when a current is applied electrons flow through the magnetron, and as they do so a magnetic field forces them to circulate, in bunches like the spokes on a wheel, across the mouth of the drilled holes. And just as the air flowing across a whistle hole causes a tone to be emitted, the frequency of which depends on the size and shape of the hole, so the electrons in the cavity magnetron oscillate at a specific radio frequency determined by the cavity dimensions. Each bunch of electrons is reproduced in the cavity walls, and the power generated there is conducted away in the form of very short waves, or microwaves.

This was the stage that Randall and Boot's research had reached by November 1939. Their work was being carried out in secret, and in what was essentially a backwater of the official research program. Even so, Oliphant had the foresight to give the two men a free hand to take the research further and see what they could come up with. Work was briefly disrupted around Christmas time when Oliphant's team relocated to a newly built facility which would come to be known as the Nuffield Research Laboratory. Over a couple of months the pair gradually got together all the equipment they needed to build a working example of the cavity magnetron. Sometimes it was crude, but nonetheless effective. Henry Boot, twelve years the junior of Randall, had excellent workshop skills—and the inspirational verve of youth. He solved a particularly tricky problem of plugging one end of the central cavity by using an old-fashioned halfpenny, a coin exactly the right size and shape for the job.

By February they had finally succeeded in building their prototype, and on February 21, 1940, they wired it up to a single car headlight on their work bench. As the switch was thrown for the first time on a cavity magnetron, the headlamp's six-watt bulb suddenly glowed with a fierce brightness; then, in an instant, burned itself out. The power generated was beyond anything they had expected.

Over the next few days they succeeded in burning out a series of bulbs of ever-increasing wattage, before moving on to more powerful neon tubes. At first the new magnetron produced a continuous wave output, but soon afterward they managed to convert it to the pulsed output that would be needed for a radar system. With dramatic results: At the required wavelength of 10 cm, this revolutionary piece of equipment could produce well over 1 kW of power. It was clear that these two men, working on a semi-freelance basis, had their hands on the Holy Grail of radar. But the next task was almost as daunting as the one they had accomplished—how to turn this test-bench marvel into an operational production model that would hold its own at the front line of war. Fortunately, work was already well under way toward establishing an industrial basis for the production of centimetric radar.

Oliphant's laboratory had been contracted by Sir Charles Wright of the CVD in the summer of 1939 to carry out research into the possibility of centimetric transmitters and receivers. At around the same time, similar contracts had been placed by the Air Ministry specifically to look into developing an airborne radar system using centimetric waves. But what made those contracts interesting was that they were placed not with universities, but with commercial companies, specialists in electrical engineering. Never before had such development work (as opposed to production) been put out to private industry. The first contract was with EMI (Electrical and Musical Industries), and the second with GEC Research Laboratories at Wembley, which had already had some success in reducing the wavelength of conventional equipment. Tizard, in his position as scientific adviser to the Air Ministry, took a leading role in negotiating these contracts, and Eddie Bowen, the man who had been responsible for the development of airborne radar at the government's research establishment at Bawdsey, was given a coordinating role. In fact in offering this role to Bowen, Tizard was throwing him something of a lifeline, at a time when the fiery Welshman needed all the friends he could get. Britain might have been in the middle of the "phony war," but for Bowen it had been anything but. A chain of events that started on the day Britain entered the war had seen Bowen's star falling fast.

Bowen had always been something of a maverick, and it had served him well under Watson-Watt, who had had the foresight to go along with Bowen's ideas in creating the airborne group in the first place. Bowen's relationship with A. P. Rowe, the man who took over from Watson-Watt as the head of Bawdsey, was very different. Watson-Watt, a scientist with a considerable track record behind him, understood the scientific mind and that bright young men would sometimes want to follow their own path. All of which was a perfectly reasonable *modus operandi* while the research establishment remained small. Rowe, his replacement, however, was not a scientist, but an administrator. His background was in the civil service, and his skill was in organization. It would be a talent that would be tested to the full in a fast-growing establishment soon to find itself operating under the pressure of war. Rowe and Bowen never really hit it off, and when war finally arrived things went from bad to worse.

When war was declared on September 3, 1939, it was universally expected that the Luftwaffe's bombers would come roaring across the English Channel. Their primary target would be London itself, but the all-too-visible radar towers of Bawdsey Manor must surely not be too far behind on the list. Indeed, just the previous month a German Zeppelin bristling with electronic listening devices had been spotted making its leisurely way up the eastern coast just off Bawdsey. Rowe correctly surmised that it was specifically looking for radio signals (and was thus the world's first electronic surveillance flight).[4] Accordingly, the day after the German invasion of Poland began he ordered that Bawdsey must be immediately evacuated. It was not a decision with which Bowen agreed. His skepticism was to some degree vindicated when war was declared the following morning, as Bawdsey's 250 staff found themselves trying to relocate an entire research establishment on a day singularly unsuited to the task, with road, rail, and air traffic in disarray and millions of people pouring out of London into the countryside.

Plans had, of course, already been laid. In an attempt to relocate as far away from the firing line as possible, Robert Watson-Watt, himself a Scotsman, had chosen the Scottish town of Dundee to be the new base of the now renamed Air Ministry Research Establishment, or AMRE. Some months earlier he had returned to his alma mater, University College, Dundee, to negotiate the use of space in the institution in the event of war. Sadly, as far as University College, Dundee was concerned this was only an agreement in principle, and one that had never been followed up, so that when Rowe and his men began to arrive with their eighty tons of crated equipment, they found that nothing had been prepared; in addition, there was precious little space to accommodate them. Two rooms were all that was available to house the steady stream of men coming up from the south, and all the very latest, state-of-the-art technology of war had to be left in the parking lot in the trucks that had brought it. To make matters worse, now that war had arrived, Rowe had begun to activate plans to draft in a further round of the country's top engineers and physicists. During that fall the staff of AMRE almost doubled.

On top of all this there was now an increased urgency to new lines of development. Principal among them was the urgent need to plug a glaring hole in the Chain Home defensive system, which was proving to be ineffective against low-flying aircraft. AMRE's scientists solved the problem by taking a short range gun-laying radar set the army had developed for their coastal batteries and adapting it so that it could scan close to the horizon. Prosaically christened Chain Home Low, the system would be installed alongside its big brother at the Chain Home stations.

Bowen's airborne radar group was shipped out to the nearest airfield some twenty miles away at Scone, near the town of Perth. Conditions here were even more unsuitable than they were at Dundee. Once again, it seemed Watson-Watt had never followed up on his original meeting. This group from AMRE was simply not expected. One hangar and a share of some office space in a still-functioning civilian airfield were all that could be made available for some of Britain's most secret war research work. Not that there was much research work going on. By now the focus had understandably shifted into the realm of practicalities. Getting the existing equipment into airplanes had a more immediate priority than further refining the design. But this was not the work that Bowen's talented team had signed up to. Sharp young research scientists to a man, they did not see themselves as aircraft fitters and mechanics. But with Fighter Command leaning heavily on them for the delivery of the airborne interception sets they had ordered, and Coastal Command desperate to get some ASV sets installed, this was what they seemed to have become. Morale was plummeting fast. It deteriorated further when, just six weeks after the move to Perth, news came through that the group was to relocate yet again.

St. Athan in South Wales is just a few miles from Swansea, the town where Eddie Bowen grew up and went to university. In 1939 it was the site of an RAF base that was the home of No. 32 Maintenance Unit, whose job was the servicing and overhaul of aircraft and their engines. Rowe saw it as the ideal location for the production-line fitting of aircraft radar, and it was to St. Athan that Bowen's airborne group was moved in the middle of October. Once again, though, their chosen home turned out to be very much less than ideal and the scientists felt sidelined and undervalued. When they arrived they found a base already overcrowded with new recruits, all of whom would need training in the most basic of maintenance tasks. Surely they could be of little use in the highly specialized installation of sophisticated new electronic equipment. Yet again the space made available to them was woefully inadequate—a large aircraft hangar with next to no protection from the winds of the approaching winter, nor any heating or even electricity. The whole of their first week was spent, not equipping fighters with the latest technological aids, but laying plywood floors, hanging makeshift canvas partitions, and installing an electrical power supply. Bowen had long felt that Rowe was a poor replacement for Watson-Watt, but now he was beginning to take it personally.

University College, Dundee was about the worst place you could possibly imagine to put a radar research establishment. In the middle of a bustling city, there were buildings all around, precious little open space and very large amounts of electrical disturbance. Rowe knew from day one that somewhere else would have to be found, and by May 1940, AMRE was on the move again, this time to the other end of the United Kingdom. Rowe needed to find "a large and fairly flat isolated area; we needed a cliff site, proximity to electric power, proximity to an aerodrome and to a town where the staff could live. We needed also to be reasonably near London yet as far as possible from enemy activity. Finally, a site was found which stretched from the charming village of Worth Matravers in Dorsetshire to the lovely cliffs of St. Albans Head, with Swanage the nearest town."[5]

Ironically the site at Worth Matravers was to remain "as far as possible from enemy activity" for only a matter of a few weeks. The convoy carrying the establishment south left Dundee on May 5, 1940, just five days before Hitler's troops invaded the Low Countries and six weeks before France would formally surrender.

In Worth Matravers at last they had the space they had so lacked in Dundee. They needed it. During the months at Dundee the AMRE team had grown substantially, now numbering some four hundred. With that growth had come the need for Rowe to appoint a deputy. Bowen, as one of the establishment's longest-serving members, a close confidant of Watson-Watt, and the leader of one of the most advanced projects so far undertaken, supposed himself to be a shoo-in for the job. Rowe, however, had other ideas. His choice for the post was a bright young scientist who had joined the establishment from Cambridge only a few months earlier. His name was W. B. (Ben) Lewis, and his appointment was just about the last straw for Eddie Bowen.

Bowen now found himself further sidelined. Safely installed in their chosen environment of Worth Matravers, Rowe and Lewis now felt prepared to pursue the goal of centimetric airborne radar.

Bowen, of course, had more or less pulled off the Herculean task of designing and building airborne radar on the old metric-wave system. But it is sometimes inadvisable to be at the leading edge of invention, the spoils often going to those who oversee the second phase, where it is refined into a practical system. And, as Rowe pointed out, at neither Scone nor St. Athan did Bowen have the proper facilities for working on centimetric applications. Whatever the rights and wrongs of the situation, once at Worth Matravers Rowe and Lewis quickly formed a new airborne radar unit under another fiercely bright young man—this time a recruit from Cambridge, P. I. Dee. The unit was specifically tasked with cracking the problem of building a workable airborne set operating on the 10 centimeter wavelength, but as yet it knew nothing of the windfall it would shortly receive in the form of the cavity magnetron.

An important program pioneered by A. P. Rowe in the early days at Worth Matravers was the so-called "Sunday Soviet," that became a key development tool of the organization. In early 1940, Rowe had had the idea of inviting senior military personnel to visit on Sundays to meet with the research engineers and scientists who made up his team. The revolutionary aspect of these meetings, particularly in the status-ridden society that was 1940s Britain, was that they were highly informal, and even the most junior staff were encouraged not just to attend but to contribute ideas. No amount of seniority, age, or experience bought you precedence. But if an idea was judged to be worth pursuing, it could be taken up there and then, because all the main decision-makers would be in the room. The informality was unprecedented, and a sense of common purpose was thus built between researchers and military decision-makers.

Some eight or nine months had now passed since the scrambled evacuation from Bawdsey on the eve of war. They had endured a chill winter of dislocation and disruption, but now, with the move south to a specially chosen and constructed site, things seemed, to Rowe, to be settling into place. Lewis, Dee, and the rest of the influx of scientists had strengthened the intellectual firepower of the organization, and in Dee's new Airborne Group, Rowe was sure that the cutting edge of development was in good hands. Bowen's outfit had become superfluous and was disbanded. Bowen himself was once again relocated, this time to Worth Matravers's home airfield some twenty miles away at Christchurch. His job, to carry on with the task of equipping aircraft with the available AI gear, was no doubt vital war work, but to Bowen it was the ultimate demotion. But he would not be alone in this position for long. In just a few weeks the same fate would befall Sir Henry Tizard, a man who held Bowen in high regard. But as yet the momentous events of the spring and summer of 1940 were only just beginning.

During the first couple of months of 1940 (before Randall and Boot's breakthrough), the GEC Lab at Wembley seemed to be making significant progress in its attempts to bring the wavelength of conventional systems down to the sort of figures that made an airborne set begin to look possible. Using a valve called a micropup they had achieved a wavelength of 50 centimeters, and were hopeful of developing a valve (the millimicropup) that would reduce it still further to around about 20 centimeters. Oliphant's group at Birmingham (the group formed while Randall and Boot were still in Ventnor) had managed to produce a narrow beam at the 10 centimeter wavelength using a Klystron; but only with equipment far too big to even think of mounting it in an aircraft. It had become clear, though, that the narrowness of the required beam meant that a way would have to be found of rotating the beam so that it could sweep the sky. It also highlighted the need for some sort of pictorial display on which multiple targets could be seen at the same time.

By early June the GEC team had managed to build a fully functional system on the roof of their laboratory at Wembley. Working at a wavelength of 25 centimeters they were getting echoes from ground objects at a range of five or six miles. For a moment it looked as though this might be the answer, until the GEC scientists' world was turned upside down by the arrival of Randall and Boot's discovery, the cavity magnetron. Still top-secret, a small team at Wembley led by Eric Megaw were given the task of turning the crude prototype from the Birmingham lab into a practical design suitable for mass production. Only then would this revolutionary discovery be able to make its mark. Over the next few weeks Megaw's team labored around the clock to perfect the new device, incorporating two significant improvements, a large diameter, oxide-coated cathode (delivered into their hands in a timely fashion by Maurice Ponte, a scientist fleeing from Paris before the Nazis arrived), and the use of a gold seal. By the end of July the experimental set on the rooftop at Wembley had been adapted so that it could use either the old micropup system at 25 centimeters, or the new cavity magnetron operating at 10 centimeters. Comparative tests produced a remarkable result—the cavity magnetron outperformed the old system by a considerable margin. In addition, because it contained its own sealed vacuum and needed no outside power source to generate one, it was a mere fraction of the size of competing systems. The cavity magnetron was tiny, generated immense power, and produced a powerful, narrow beam at the desired wavelength of 10 centimeters. The search for the Holy Grail was over. All that remained to be done now was to find ways to exploit it, and, above all, the means to manufacture it in the quantities and at the speed which the rapidly deteriorating state of the war demanded. Fortunately, work was already afoot which might, just possibly, facilitate this.

NINE

Three Thousand Miles from War

PRESIDENT ROOSEVELT WAS NOT ALONE in being influenced by the fast-moving events in Europe. Many of the leading scientific academics in the United States had been following the developing situation closely, and many now came to the firm conclusion that America's entry into the war was only a matter of time. They, more than most, were keenly aware that this would be a technological war, and that those countries who failed to prepare properly by drawing their finest scientific minds into the war effort at the earliest possible moment, would pay a high price when conflict was joined.

In 1938 Vannevar Bush had joined the National Advisory Committee for Aeronautics. NACA had been established by Congress in 1915 "to supervise and direct the Scientific Study of the problems of flight," and had, for nearly twenty-five years, overseen research work that had been of incalculable benefit to the development of both civil and military aircraft in United States. Bush came from an old New England family. He was professor of electrical engineering at MIT, and an inventor with invaluable experience of the business world. His appointment to NACA was a vital step on the road to becoming America's most powerful wartime scientific administrator. Such were his talents that by the following year, 1939, shortly after he resigned his vice presidency of MIT to become president of the Carnegie Institution of Washington, he became chairman of NACA.

His rise to the chairmanship was fortuitous, as it coincided with the outbreak of war in Europe. It was clear to Bush from the start that the United States was in need of an organization with overall control of research and development for the application of science in wartime, and he quickly set about canvassing opinion from some of the most powerful academics in the country. Principal among them was his former boss, Karl T. Compton, president of MIT. He also consulted his colleagues at NACA together with President James B. Conant of Harvard, and Frank B. Jewett, the president of the Bell Telephone Laboratories and of the National Academy of Sciences. In May 1940, John Victory, the secretary of NACA drafted under Bush's direction an Act of Congress to set up an organization "to coordinate, supervise, and conduct scientific research on the problems underlying the development, production, and use of mechanisms and devices of warfare, except scientific research on the problems of flight."[1]

The proposed body, the National Defense Research Committee, would consist of not more than twelve members, appointed by the president himself, and would include one member each from the War and Navy departments, as well as the National Academy of Sciences. It was to have sweeping powers to contract educational institutions, individuals, and industry to produce scientific studies and reports. Its members would serve without remuneration. While they would work closely with the military they would, crucially, be independent. Early in June 1940, Bush was able to discuss the idea with Roosevelt who immediately saw its value and agreed to write to the Secretaries of War and the Navy, and to the president of the National Academy of Sciences, requesting their full support and cooperation for the new agency. Now some powerful names were added to the roster of potential members. To Compton, Conant, and Jewett were added Conway P. Coe, the Commissioner of Patents, and Richard C. Tolman of the California Institute of Technology, who, convinced that the United States would soon be drawn into the war, had recently traveled to Washington to make his services available in whatever way was required. On June 27, Roosevelt signed the order approving the NDRC, making Vannevar Bush its chairman, and stipulating that the War and Navy departments send senior and distinguished officers to represent them on the committee. It took him just ten minutes to approve the creation of this ground-breaking organization which was designed to bring together three communities so far united by little more than their suspicion of each other—scientists, industry, and the military. Their common task now would be "to correlate and support scientific research on the mechanisms and devices of warfare."[2] With the words "OK—FDR," the president authorized the start of a movement that would shape America's preparations for war. Within a few days the Army and Navy, who already appreciated that the demand for military research might soon outstrip their capacity to meet it, came up with lists of projects that they would like the new agency to take on, and the membership of the committee was soon completed with the addition of Brigadier General George V. Strong and Rear Admiral Harold G. Bowen as the service representatives.

But Roosevelt knew he could achieve nothing by openly swimming against the tide of public opinion—except possibly his own downfall. Time and again he publicly stated that he had no intention of sending American soldiers abroad to fight in someone else's war. It was a claim he would continue to make publicly right up to and including his campaign for a third presidential term in the fall of 1940. Wendell Willkie, Roosevelt's opponent, a former Democrat and a firm supporter of Britain, was no isolationist at heart, but by the last few weeks of campaigning he had aligned himself strongly with the antiwar faction. Willkie was now presenting himself as the peace candidate. In speeches in Chicago and Baltimore late in October he painted Roosevelt as intent on involving the U.S. in Europe's war, telling his audience that Roosevelt's promise "to keep our boys out of foreign wars is no better than his promise to balance the budget, they're almost on the transports!" and insisting, "If you re-elect him you may expect war in April 1941." Roosevelt, for his part, and notwithstanding his private convictions, again restated his public position in a speech at Boston just as Willkie was speaking in Baltimore. "I have said this before," he told the "mothers and fathers" of America, "but I shall say it again and again and again. Your boys are not going to be sent into any foreign wars." He omitted to add the rider, explicitly stated in Democratic policy documents, "except in case of attack." When questioned privately by one of his aides about the omission, he defended his verbal sleight of hand by pointing out that if the United States were attacked then it would no longer be a "foreign war."

All this came at the end of several years during which Roosevelt had been fighting a long battle to overturn the provisions of the Neutrality Acts of the mid-1930s designed to ensure that America would not be sucked into another war in Europe. The midterm elections of November 1938 had brought many more Republicans into the U.S. House of Representatives, who, when combined with conservative Democrats, could easily vote down the president's "unneutral" policies.[3] Roosevelt's chances of loosening the grip of the Neutrality Acts were looking slim.

For decades Roosevelt had followed the orthodoxy of a strong navy forming the linchpin of the United States defences. It made sense, after all, with several thousand miles of ocean between the U.S. and its nearest potential aggressor. But now the world contained the new and terrifying threat of aerial bombardment. Roosevelt was quick to see that even those thousands of miles of ocean might not be enough to protect America from the newest aircraft on the drawing boards of the top engineers and designers. There had long been a fear in the U.S. that fascism might transport itself across the Southern Atlantic to Latin America. German trade with that continent was on the increase, and there were substantial numbers of German immigrants in Brazil and Argentina. If the Nazis were to establish a base in, say, Brazil, followed by the seizure of the Azores or the Cape Verde islands, it would only be a matter of time before Nazi bombers were stationed within range of America's southern states. To guard against such an eventuality Roosevelt decided in the autumn of 1938 to station a small part of his navy in the Atlantic, and plans were laid for the annual fleet maneuvers during February 1939 to take place off the East Coast.

As the summer of 1939 arrived the situation in Asia was looking increasingly dangerous. With war in Europe now looking inevitable the British were keen to avoid getting embroiled in any conflict in the Far East. Early in the summer, four Chinese nationalists had taken refuge in the British enclave at Tianjin, a major Chinese port to the southeast of Beijing. The British had refused a request from the local Japanese Army commander to hand them over, and as a consequence he blockaded the settlement. Britain was in danger of being sucked into a far eastern conflict she could well do without, and on July 24 the British ambassador in Japan signed an agreement bringing this potential crisis to an end by acknowledging Japan's "special [security] requirements" in China. The the U.S. State Department regarded this as setting a dangerous precedent, and on July 26, Roosevelt announced that Japan was to be given six months' notice of America's decision to abrogate the U.S. trade treaty with Japan—a treaty that dated back to 1911. It was a step that opened the door to the possible imposition of trade sanctions from the beginning of 1940. The temperature of relations between the United States and Japan had just dropped a degree or two.

Russia now stepped into the Asian limelight. Since the autumn of 1938, Japan had been in control of most of northeastern China as far south as Shanghai. Six years previously, Japanese expansionism had reached the borders of Russia when it annexed Manchuria, a mineral-rich region on Russia's eastern doorstep. Access to new sources of raw materials were essential to Japan's intended development as a major industrial power. For months there had been a series of skirmishes along the border between Manchuria and Outer Mongolia, now a Russian "protectorate" following the conclusion in 1936 of a ten-year Mongolian–Soviet Treaty of Friendship. On August 20, 1939, these border skirmishes exploded into full-scale war when Soviet forces launched a preemptive strike against the Japanese. Three days later this unexpected incursion suddenly made sense, with the public announcement that Russia and Nazi Germany had signed a non-aggression pact. Japan, which had for some time been seeking closer ties with Germany, was suddenly left high and dry. Japan's aggressive, expansionist foreign policy had been dealt a significant blow. For a year or so it would lie dormant, before returning with a vengeance in the summer of 1940.

The Nazi–Soviet Pact may have stopped Japan in its tracks, but it opened the door to fundamental movements in Europe, sealing the fate of Poland and plunging much of Europe into war. It was the signal for Roosevelt to recall Congress to readdress the Neutrality Acts now that the expected war had finally arrived. But not before he had set out his position to the people on the evening of that Sunday, September 3. Neutrality would henceforth have a new meaning: "This nation will remain a neutral nation, but I cannot ask that every American remain neutral in thought as well. . . . Even a neutral cannot be asked to close his mind or his conscience."[5] In August 1914 President Woodrow Wilson had asked Americans to be "impartial in thought, as well as action."[6] Twenty-five years on, this president clearly felt neutrality in the ideological battle was not an option.

The revised legislation now put before Congress differed in one key aspect from the existing situation. The mandatory arms embargo was to be replaced by a policy of strict cash and carry. If you could pay, you could trade with America. Opponents of the new legislation, which was to Britain and France's advantage, with their large reserves of foreign exchange and big merchant marine fleets, argued that even passively favoring one side of the conflict in this way would potentially make the United States a target for the other. For a few weeks the debate galvanized the nation, until Roosevelt won the day, and on November 4, 1939, the president was able to sign the new Neutrality Act into law. It had taken the outbreak of war in Europe to do it, but Roosevelt had finally achieved the position he sought. His goal, quite simply, was to keep America out of this war by supporting the ability of the British and French to fight it.

Revision of the Neutrality Acts so as to allow cash sales to Britain and France would not only support his "first line of American defense," but would also start to build a strong arms industry in the U.S., particularly in the aviation sector, capable of arming America quickly if and when the moment came. There was more than a hint of the New Deal about this policy. Roosevelt told his aides, there was an election in the offing, a little more than a year away, and "Foreign orders mean prosperity in this country, and we can't elect the Democratic Party unless we get prosperity."[7]

But surprisingly, now that the legislative block had been moved aside, no orders came. The British, it seemed, were reckoning on a long war—maybe as long as three years—and they were at pains to preserve their currency reserves. In fact, Britain even started to cut back on some of its commercial trade with America, shifting purchasing of goods such as tobacco and fruit to the Mediterranean in a desperate attempt to prevent Italy, Greece, and Turkey from falling into the German orbit. Roosevelt himself had done much to lock British and French policy together by encouraging them to establish a joint purchasing Commission and to channel their orders through the U.S. Treasury. And there was a major fiscal shift between this war and the last. This time American law prohibited Britain and France from raising loans in the United States. There was no way around it: buying American planes would mean a major outflow of the currency reserves the British and French would need to tide them through a long war. It was a policy fraught with danger.

The European Allies had every reason to suppose that this war might be a long one, because in some respects it had not yet even started. Churchill's new posting as Lord of the Admiralty was something of a galvanising influence on the Anglo-French Supreme War Council, which on March 28 took two key decisions, first to begin mining Norwegian waters, a policy that was fatally delayed and, in the event, pre-empted by the German invasion of April 8, and, secondly, at last to place an order for the purchase of 4,600 American planes. So, in the spring of 1940 the logjam of the Phony War was broken by the opening of the Norwegian campaign and by the European Allies' placing their first major order for matériel with the United States—the first of what would soon become a major stream.

In fact the president himself would be compelled to revise his entire mindset. The journalist Joseph P. Lash, a close friend of the first lady, Eleanor Roosevelt, records that "until the end of the Norwegian campaign, the president's formula for the restraint of Nazism in Europe had been a 'trinity of French land power, British sea power and American industrial power.'"[8] French land power had now been swept away, and the British Navy could be about to fall into the hands of the Germans, which would leave American industry with nothing to drive.

At this moment perhaps the most significant thing to underpin the continuing fight against fascism was the burgeoning relationship between Roosevelt and Churchill. It was a powerful and complex bond that developed between two very different men, but one which held that transatlantic lifeline open at this crucial moment. Churchill's demeanour in the face of adversity did as much to stiffen the resolve of Roosevelt as it did that of the British people. The American president was mightily impressed by Churchill's "Dunkirk" speech on June 4, when he pledged the nation's continuing resistance in the face of the threat of invasion in the most trenchant and inspiring of words, "we shall defend our Island, whatever the cost may be. We shall fight on the beaches, we shall fight on the landing grounds, we shall fight in the fields and in the streets, we shall fight in the hills," and finishing by vowing "we shall never surrender." Throughout that summer Roosevelt worried that Britain's will to resist would collapse. But it was Churchill's fortitude that overcame these doubts and convinced the president that Britain could hold the Nazis at bay and prevent Hitler gaining control of the Atlantic.[9]

And, for his part, Churchill had long looked across the Atlantic for the wellspring of his political ideals. Sir Isaiah Berlin wrote eloquently shortly after the war about Churchill's views of the nations ranged against each other in World War II:

> [He] has never hated Germany as such: Germany is a great, historically hallowed State; the Germans are a great histrionic race and as such occupy a proportionate amount of space in Mr Churchill's world picture. He denounced the Prussians in the First World War, and the Nazis in the Second; the Germans scarcely at all. He has always entertained a glowing vision of France and her culture, and has unalterably advocated the necessity of Anglo-French collaboration. He has always looked on the Russians as a formless, quasi-Asiatic mass beyond the walls of European civilisation. His belief in and predilection for American democracy are the foundation of his political outlook.[10]

And in this darkest of hours for his country, Churchill very much liked what he saw in the White House. According to Isaiah Berlin:

"he trusted Roosevelt utterly, convinced 'that he would give up life itself, to say nothing about office, for the cause of world freedom now in such awful peril.'"[11]

Churchill's resolution and decisiveness were underpinned by the fact that he headed a national government, drawing on talents from all parties, uniting the nation behind a single cause. Now Roosevelt himself began to shape his administration along similar lines. His appointment of Knox and Stimson, two veteran and senior Republicans, signified that his platform in November, though Democratic in name, would be bipartisan in spirit and character. What of the Democrats' defining characteristic through the thirties, their commitment to public works and the New Deal as a means of lifting the country out of the depression? Ardent New Dealers were quick to spot what was going on here. They immediately saw a sinister conjunction between Republicans and senior industrialists (including Stimson) being brought into government, and the NDRC with its "powers to contract . . . industry." A little over eighteen months had passed since Roosevelt had briefed officials on his grandiose plans to build government-owned factories capable of turning out 10,000 planes, and taking vast armies of workers off relief, the whole project to be handled by the Works Progress Administration. A dramatic shift in rearmament policy was being signaled here, from public to private factories, and the New Dealers knew it. The word "betrayal" even entered their vocabulary. For Roosevelt it must have been a hard road to take, but it was vital. Civilian scientists and industry were shortly to become the great engine of innovation and manufacture of new weapons and of the vast array of technological breakthroughs.

Henry Tizard, left, and Frederick Lindemann were the two leading advocates of a scientific response to the threat of aerial bombing, particularly as aircraft continued to increase in speed, range, and armament following World War I. (*British Official*)

Robert Watson-Watt, left, suggested that radio waves could be used to detect aircraft and led the scientific team that developed practical radar for the British. Archibald Vivian Hill, right, received a Nobel Prize in 1922 for his work on the production of heat in muscles. He was sent to the United States in 1940 at Tizard's request to help lay the foundation for an American and British scientific exchange. (*British Official*)

Top, Bawdsey Manor, a country house located along the Suffolk coast, became the main research station for work on radar in Britain before World War II. (*Author*) Bottom, Britain's radar research was moved to Worth Matravers in Dorset once the war began in order to be further away from enemy action. Opposite: top left, a Chain Home Low station was an adaptation to provide early warning against low-flying aircraft; top right, an original Chain Home station that provided Britain's first defense against incoming aircraft by identifying them at great range; middle, left, an operator at a Chain Home station viewing the monitor and equipped with a "hands-free" telephone; bottom, the outline of the River Severn at 1000 feet, revealed by H2X radar, possibly the first-ever X-band radar image. (*British Official*)

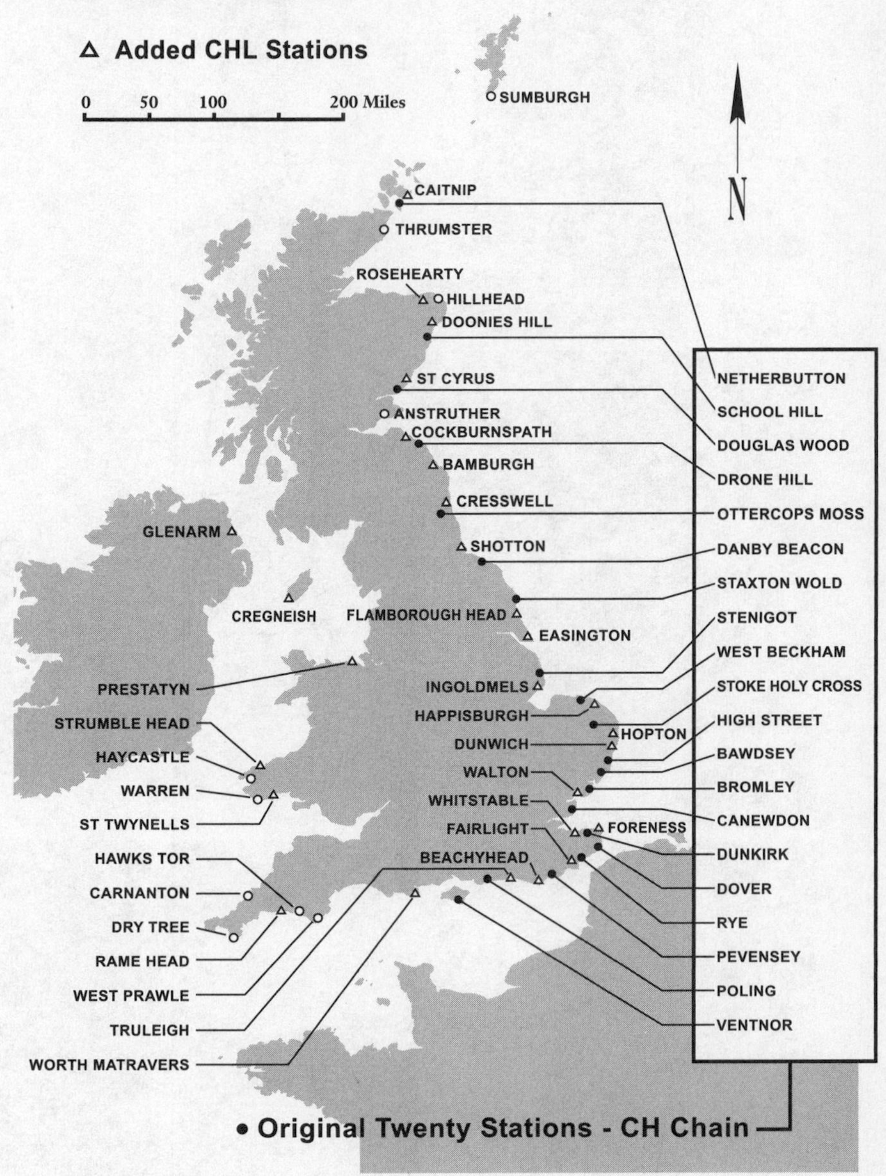

Chain Home Stations and Chain Home Low Stations.

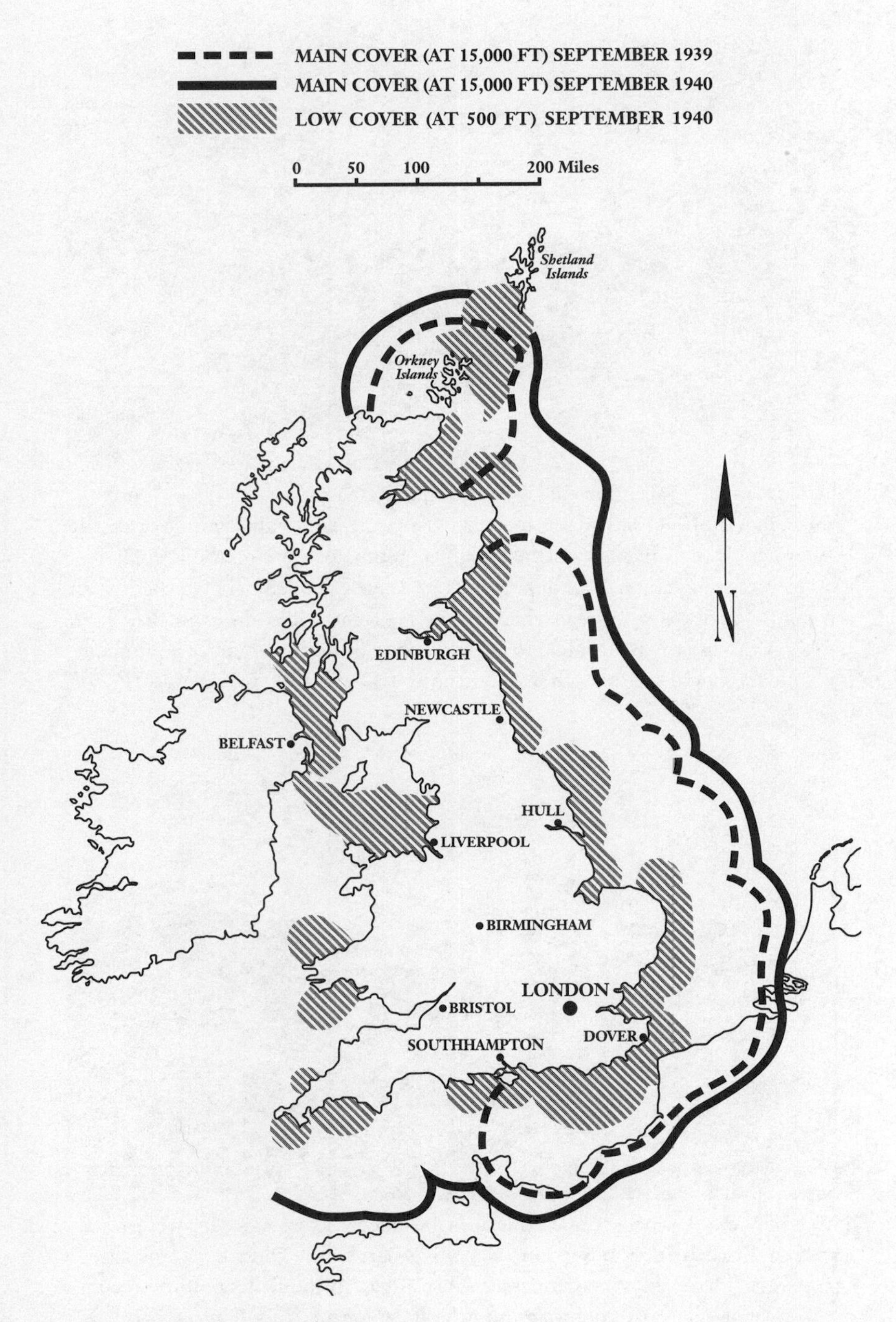

Chain Home Station radar coverage.

The successful evacuation of British Expeditionary Forces from Dunkirk, France, in May 1940, was an unexpected twist in the early fighting in World War II. Germany swept through northern France, ultimately forcing the French military to surrender, and had an opportunity to destroy the remnants of the British army at Dunkirk, but instead checked their advance, allowing the British to escape. Although the British had to abandon an enormous amount of equipment, the soldiers would have another opportunity to fight later in the war. (*British Official*)

Three of the old World War I American flushdeck destroyers that were transferred to Great Britain in September 1940 as part of the "destroyers for bases" agreement. These ships were an essential stop-gap in the British effort to combat U-boat attacks on shipping in the Atlantic. (*Naval History Center*)

Vannevar Bush, seated, confers with Assistant Secretary of Commerce, Richard C. Patterson, Jr., during Bush's congressional testimony in early 1939, where he urged the U.S. government to encourage scientific and industrial research in the event the country found itself in a "major difficulty." (*Library of Congress*)

Members of the Tizard Mission and British attachés. In the front row on the left is Ralph Fowler, scientific attaché to Canada, the mission stenographer, Miss Geary, may be the woman in the center. In the back row, from left to right is Brigadier Charles Lindemann (Frederick Lindemann's brother), the British Army attaché in Washington, Sir Henry Tizard, and George Pirie, the British air attaché in Washington. (*British Official*)

A German Heinkel He-111 bomber shown at the beginning of World War II. This aircraft was first used during the Spanish Civil War and made up the bulk of German bombers attacking Great Britain in 1940 during the Battle of Britain. It was the development of this kind of aircraft in the 1930s that alerted British strategists to the vulnerability of their defenses. (*Bundesarchiv*)

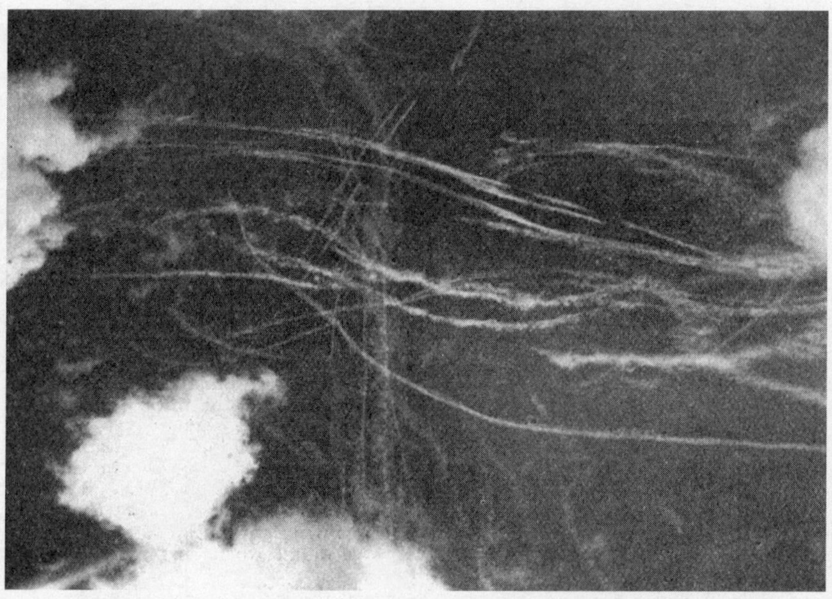

A tangle of contrails left by swirling British and German fighter aircraft in a dog-fight over Kent in September 1940. This is graphic example of the importance of IFF (Identification, Friend or Foe) used by the British planes to distinguish friendly from enemy planes. British IFF technology was swiftly adopted by the American Army Air Forces when it was first disclosed. (*British Official*)

During the Battle of Britain, German strategy changed from attacking airfields and other military targets to bombing cities and towns. Known as the Blitz, bombs rained down on both commercial and residential districts causing raging fires and immense piles of rubble from ruined buildings. These photographs show three different attacks on London: the John Lewis & Co. department store (upper left), among buildings destroyed in the first bombing of central London; a raging fire in East London as a result of a nighttime raid; and a mountain of bricks and other debris from houses destroyed on Hallam Street. (*British Official*)

A German submarine being attacked by a British Coastal Command B-24. Radar carried in aircraft was the principal weapon that defeated the U-boats. (*British Official*)

An American destroyer transfered to Britain, left, is fitted with air search radar atop its mast and a microwave surface search set in the lantern-like structure above the bridge. Microwave radar was essential for convoy duty both to identify ship positions and to locate enemy submarines. An SCR-720 microwave radar, right, in the nose of an American P-61 Black Widow night fighter. Airborne radar was a key weapon in the defeat of Germany and Japan in World War II. (*National Archives*)

TEN

The Mission Takes Shape

WITH THE ARRIVAL OF THE MOMENTOUS MONTH of May 1940, A. V. Hill was still in the United States trying to advocate for an open scientific exchange. But Churchill's rise to power shifted the ground suddenly and dramatically under Tizard and Hill and their radical proposal. The need to enlist American help had now become urgent. They must also have known, though, that they themselves were in a position of some jeopardy, since now that Churchill was surrounding himself with key allies in all the top posts, so the star of Lindemann, with whom they had so recently clashed, would also be in the ascendant. Tizard's days of influence were numbered. In the turmoil of late May it became apparently impossible to get a decision from Whitehall and Downing Street on Hill's advice that the time was right for Britain to give away its most precious secrets. But Tizard was very much the professional middle-class English scientist, not given to histrionics, and, at times, a master of phlegmatic understatement. His papers record that on May 31 he wrote Hill that: "everybody was so hot and bothered on the Air Staff that I did not know what had happened about his suggestion that we should give out secret information on certain matters to the United States Government."[1]

The two men decided between them that the best chance of pushing their plan through to fruition was for Hill to come home immediately and make the case in person. It may well be that Tizard thought that his own ability to persuade those at the highest level of government was considerably diminished now that Lindemann had the ear of the new prime minister. Better, then, to bring Hill straight back in the hope that they would listen to the man who had recently been on the spot. Accordingly, Hill booked his return passage across the Atlantic. Such was the pace of events in Europe that he left the United States just as the last troops were being plucked from the beaches of Dunkirk on June 4, yet by the time he arrived at Liverpool ten days later, the German Army was poised to enter Paris in triumph; he later wrote laconically to Tizard that he must have traveled "on a rather slow ship." He went on to observe, "They didn't seem to take much notice of what I said till after I got back. . . . After that, being an MP, I could apply a certain nuisance value."[2]

When Hill arrived back in London Churchill had been prime minister for a little over a month. He had had to construct a new government, a national government balancing the interests and skills of three political parties. He had made several perilous flights across the Channel to negotiate with the French. He had supervised plans for resisting a German invasion, and lately had been faced with a new foe entering the war in the shape of Italy. Churchill had been a busy man.

Not so busy, though, that Hill and Tizard's proposal had been neglected. In fact it had been one of the first things Churchill considered on taking office. On May 11, one day after arriving in Downing Street, his Assistant Parliamentary Secretary wrote to John Balfour, head of the Foreign Office American Department, revealing Churchill not to be in favor of giving away information on RDF (radar). A note in Tizard's archives records the position on May 11: "Winston Churchill not in favour of the interchange of information 'unless we can get some very definite advantages in return.'"[3]

On May 18, the Chiefs of Staff Committee met. Tizard had spent much of the previous fortnight trying to lay to rest the Admiralty's worries that information might pass from America to Germany. The Air Ministry's Directorate of Military Intelligence assured him that they had no evidence of such leaks. Furthermore, the Admiralty itself had come up with no actual proof of any breaches of American security. Tizard himself summarized the matter tersely: "Are we to assume that our intelligence service is quite incapable of extracting information from America and that the German intelligence service finds it quite easy?"[4] By the time of this meeting, whether as a result of Tizard's work or not, the Navy's position had substantially altered. Admiral Sir Dudley Pound, the First Sea Lord, now came down in favor of the mission, adding that the latest information indicated that the Americans were doing some useful work on microwave radio tubes. For the RAF, Sir Cyril Newall also favored sending a mission. Only the Army, in the form of General Sir Edmund Ironside, remained unpersuaded. The Americans, he felt, would have nothing to offer the British Army. Pound was not to be put off. He decided to refer the matter up to the politicians and sent a note to the new First Lord of the Admiralty, A. V. Alexander.

Pound's note had the desired effect and two days later, on May 20, the First Lord wrote to the prime minister summarizing the Admiralty's position. Not only were they now in favor, but they now took the radical stance that there was nothing to be gained by insisting on reciprocation. "I consider it important, if the offer is made, there should be no appearance on our part of attempting to bargain, and I would be prepared to release securities without restraint, including ASDICs, RDF and measures for countering magnetic mines."[5]

Now it seemed the military had decided to throw their weight behind the idea yet Tizard's papers record that on the following day (May 21): "Winston Churchill still had doubts."[6] Agreement at the highest level was clearly going to be difficult to come by.

Amidst all the unsubstantiated claims of intelligence leaking from the U.S. Navy, it suddenly became apparent that there had indeed been a breach of security, though from an entirely different source.

On May 20, MI5, the British domestic intelligence service, in the company of the Metropolitan Police, raided the London apartment of a twenty-nine-year-old American clerk at the U.S. Embassy in London who encoded and decoded diplomatic messages prior to their transmission via cable. His name was Tyler Kent, and MI5 had been tailing him ever since he had been posted to London in October of the previous year. Before that he had worked at the American Embassy in Moscow, from where he had been moved to London after coming under suspicion of spying for the Russians. No direct evidence had been found implicating him in this, so presumably it was felt he could do less damage in the embassy of a country on friendly terms with America. Unfortunately, the personal line of communication between Churchill and Roosevelt that had proven essential in maintaining the relationship between their two countries had been handled through the embassy in London, and one of the men handling this traffic was Tyler Kent. After his transfer from Moscow he was soon seen in the company of Ludwig Matthias, a suspected German agent. He also struck up a friendship with Anna Wolkoff, whose father was a former naval attaché of Imperial Russia in London, and whose mother had once been in the service of the Tsar. Through Wolkoff he met a young woman called Irene Danishewsky. Her husband was a frequent visitor to the Soviet Union and they became lovers. MI5 had a lot to go on.

Two days before Kent's arrest, Ambassador Joseph Kennedy had been presented with the evidence of his activities. Kennedy was regarded by most of the British political elite as no friend of Britain. He did little to disguise his conviction that, sooner or later, Britain would cave in to the Nazis. William Bullitt, the U.S. ambassador in Moscow (and Tyler Kent's former boss) later revealed that the Foreign Office dossier on Kennedy ended with the words: "Mr. Kennedy is a very foul specimen of double-cross defeatist. He thinks of nothing but his own pocket. I hope that this war will at least see the elimination of this type."[7]

Nevertheless, whatever his personal beliefs, Kennedy was appalled that his embassy seemed to be housing a spy. He agreed immediately to waive Kent's diplomatic immunity so that Kent could be arrested. MI5's trawl of Kent's apartment turned up almost two thousand official documents. As well as Churchill's cables, there was a book containing the names of people under surveillance by Special Branch and MI5 and a set of keys to the U.S. embassy code room.

Kent's treachery could have done enormous damage to relations between the governments of Britain and the United States, and an immediate consequence was that, after May 20, no more cables passed between Churchill and Roosevelt until the middle of June. However, the damage was limited by both sides agreeing to keep the whole thing under wraps. No information was made public until May 31, when the U.S. State Department announced that Kent had been fired and "detained by order of the Home Secretary." The fact that he had been arrested and charged under Britain's Official Secrets Act was not mentioned. In Britain the story was kept well away from the front pages. The British felt that allowing the affair to become public knowledge would have seriously damaged Anglo-American relations at this most delicate of moments. The Americans, for their own part, made no attempt to have Kent extradited to the United States. The Roosevelt administration was happy to keep it quiet, because the cables that had passed between Churchill and Roosevelt clearly showed that the president was actively looking at ways to get around the Neutrality Acts in order to help Britain survive the German onslaught. Moreover, it might have completely undermined Roosevelt's bid for reelection before he had even decided to make it. The potential damage was eloquently expressed at the time by one of the few people in the know, the Democrat Breckinridge Long, who wrote in his diary: "A dispatch from London catalogues the papers found in Kent's rooms. They are a complete history of our diplomatic correspondence since 1938. It is appalling. . . . It means not only that our codes are cracked a dozen ways but that our every diplomatic maneuver was exposed to Germany and Russia. It is a terrible blow—almost a major catastrophe."[8]

Kent was tried and convicted later that year at London's Old Bailey in conditions of the utmost secrecy. He was sentenced to seven years' penal servitude.

Returning from America on June 13, A. V. Hill was quite sure that the American scientific community were keen to get involved in the war effort. He had become convinced that Robert Watson-Watt had been completely wrong in thinking that the Americans had nothing to offer on the radar front. On the contrary, he felt that some significant developments were being worked on. What they did not have as yet though, in contrast to the British, was a centralized research establishment working flat out under the pressures of war. But he knew that there were plans afoot with Vannevar Bush to create the NDRC and that there was a willingness to make available the productive power of the giant American technology companies. All that remained was for agreement to be achieved at the highest possible level that a scientific and technological exchange should go ahead.

Churchill's last response to the idea had been decidedly lukewarm. He, and others, were anxious not to give away Britain's radar secrets cheaply. But with France invaded and British troops pinned down at Dunkirk, the situation had deteriorated rapidly since then; by the time Hill returned in mid-June it was getting still worse. Time, it seemed, might be ebbing away, and yet this state of jeopardy could be the key with which to unlock the political stalemate. On June 18, Hill completed the report of his mission. He wrote appreciatively of American technical capabilities, saying that they had "a strong desire" to help. But he warned that the American services were just as desperate to keep their latest technology under wraps as were the British. He described it as a "fetish of secrecy." He believed that only through "a frank offer to exchange information and experience" would the Americans be persuaded to agree to share their secrets. Hill stressed that this "frank offer" should be without conditions, and he hoped that in return the United States would lift all "restrictions on information or purchases."[9]

Hill circulated his reports widely. Lindemann was a key recipient, of course, but someone who took a particular interest in them was Archibald Sinclair, the Secretary of State for Air, who asked Hill to come along for a meeting. Sinclair listened politely, then arranged a further meeting with Air Marshal Joubert, who, now that Tizard had resigned as of June 21, had taken overall responsibility for radar and was keen to promote a scientific mission. Joubert had previously been unable to persuade Churchill to give the go-ahead for the mission, but now, as he met with Churchill's key ministers in the war cabinet, he had, in addition to Hill's report and Tizard's of early May, a new and persuasive argument up his sleeve.

Despite Tizard's efforts to assuage their fear, the Admiralty had long been afraid that revealing Britain's radar secrets to the Americans could result in their being leaked to Germany. Now, it seemed, Air Ministry officials had been of the opinion that Germany might well have its own defensive radar system, and now the Intelligence Branch reported that a German prisoner "stated that the Germans now have a coastal chain of stations which detect British aircraft by reflection methods." If the Germans already had defensive radar, surely the danger of a leak now paled into insignificance alongside the value of the rapid development and mass production of Britain's own new technology. Sinclair found the argument persuasive, and agreed to approach Churchill again. On June 25, Sinclair, a key member of the inner War Cabinet, wrote to Prime Minister Churchill suggesting that the time was now right for Hill's "frank offer to exchange information and experience," stressing Hill's conclusions that the Americans had much to offer and adding the new information that the Germans might be on to much of it anyway.

This latest news may well have been the tipping point for Churchill, coming, as it did, just four days after the young scientist R. V. Jones had convinced the cabinet of the existence of the Germans' *Knickebein* system for directing bombers. Perhaps there was now little to lose and a great deal to gain, and so on June 30, Churchill told Sinclair that a technical mission could be sent to the United States "provided the specific secrets and items of exchange are reported beforehand."

In Washington Lord Lothian must have seen Churchill's decision as something of a victory and a personal vindication. More than two months had passed since his April telegram advocating an urgent technical exchange. He had urged Hill, on his return to Britain in early June, to lobby hard for a technical mission to the United States. He had become frustrated when Hill's lobbying failed to produce immediate results, so much so that on June 27, he had cabled the Foreign Office to express his dismay that his April telegram had not been acted upon. Though he had not yet discussed the matter personally with Roosevelt, he felt sure, he said, that any requests for technical exchange would now be falling on fertile ground. He advised that the British could expect support not just at the presidential level, but also from the military and commercial companies. The text of this telegram was not made public outside the Foreign Office until five days later, on July 1, but it is hard to imagine that the personal views of the ambassador to the United States on such a significant matter would not have found their way to the ear of Churchill prior to his decision on June 30.

In fact the general climate was now so supportive that it took barely a week to agree on the text of a proposal, via Lothian, to be dispatched to the White House on July 8. The basic thrust of that note was drafted in London by the Air Ministry and approved by both the Admiralty and the Foreign Office. The text may have originated in London, but the tactic was very much from Lothian, who judged that the White House itself was the most favorable territory, and that by sending the offer directly to the president, any lingering opposition to the plan on the part of the State Department and the Army or Navy would be seriously undermined. Lothian, after all, was picked for the job precisely because he understood Americans better than most. It set out the position in starkly simple terms:

> Should you approve the exchange of information, it has been suggested by my Government that, in order to avoid any risk of the information reaching our enemy, a small secret British Mission consisting of two or three service officers and civilian scientists should be despatched immediately to this country to enter into discussions with Army and Navy experts. This mission should, I suggest, bring with them full details of all new technical developments, especially in the radio field, which have been successfully used or experimented with during the last nine months. These might include our method of detecting the approach of enemy aircraft at considerable distances, which has proved so successful; the use of short-waves to enable our own aircraft to identify enemy aircraft, and the application of such short waves to anti-aircraft gunnery for firing at aircraft which are concealed by clouds or

darkness. We for our part are probably more anxious to be permitted to employ the full resources of the radio industry in this country with a view to obtaining the greatest power possible for the emission of ultra short waves than anything else.
Lothian

This extraordinary note, with its cryptic references to handing over the secrets of the use of "short waves," was an offer whose generosity the Americans were hardly in a position to appreciate, knowing as yet nothing of the cavity magnetron that would so change the technology of war. Roosevelt's response would be crucial to the future path of the conflict.

As the president faced a profoundly difficult decision, what he needed was first-hand information. His new Navy Secretary, Frank Knox, had already arranged for a close and trusted friend, William J. Donovan, to travel to England to assess Britain's chances of holding out. Now Roosevelt decided to raise the ante by having Donovan travel as his own personal representative. An Irish-Catholic Republican and hero of World War I, Donovan carried substantial weight with influential individuals opposed to Roosevelt. Once in Britain he was afforded access to everything and everybody he wished to see, "from King and Churchill down," according to the U.S. military attaché in London. By the time of his departure at the beginning of August, Donovan was giving odds of "60/40 that the British will beat off the German attack. I will say a little better, say 2 to 1, barring some magical secret weapon."[10] Once back in Washington he briefed cabinet officers, congressmen, and the president himself that Britain was very much worth helping.

Predictably, perhaps, there were objections raised from certain quarters in the military—from the War Plans Division of the Army, and also from the director of the Naval Research Laboratory, both of whom sought to pour cold water on the idea that the British might have any technology superior to their own. But the proposal met with unqualified support from America's scientific community. One of Vannevar Bush's first acts as head of the newly formed National Defense Research Committee was to arrange a clandestine meeting with the British. On July 8, he met privately with George Pirie, the British air attaché in Washington, and Brigadier Charles Lindemann, his Army counterpart (and Frederick Lindemann's brother). At Bush's suggestion the meeting took place at Pirie's home, where he was able to offer advice on how Lothian's proposal should be presented to the administration. Just as useful, however, was the fact that this meeting was the springboard from which the NDRC's close working relationship with the British technical mission would be launched.

Roosevelt was now surrounded by men sympathetic to Britain's needs, men who believed that offering practical support to Britain at this time was in America's best defensive interests. Henry Stimson, the new Secretary of War, and his Navy counterpart, Frank Knox, both fell firmly behind the British proposal at a cabinet meeting on July 11. A week later news of this cabinet endorsement was conveyed confidentially to Lord Lothian by the State Department, and on July 22, he received confirmation directly from the White House. Roosevelt, he was told, "would be glad if [the] special mission could come as soon as possible." Formal approval also arrived that day by the War and Navy departments of "full and free interchange of technical information," subject to two conditions.[11] The first was that equipment would only be supplied to the British if it could be shown to fall outside the requirement of America's military procurement. The second condition meant that the "full and free interchange of technical information" was not to be so "full" after all, because certain items of a "highly sensitive" nature were to be excluded from the exchange. Included among these, of course, was the Norden bombsight.

Nonetheless the path had now been cleared for the mission to go ahead, and on July 29, the State Department sent its formal approval of the mission to Lord Lothian at the British Embassy. After many months something quite astonishing had been achieved—a technical exchange of secret scientific military information between two sovereign states, only one of which was as yet at war. Time now was the only enemy, and things would have to happen fast.

These negotiations were taking place in the context of a war that was becoming ever more difficult to ignore—July 2 saw the sinking of the British merchant ship *Arandora Star* off the coast of Ireland by a U-boat. Of the 1,200 people aboard 800 were drowned, among them "enemy aliens" being deported from Britain for internment abroad. Two days later the Luftwaffe attacked a convoy in the English Channel south of Portland. Stuka dive-bombers sank five out of the nine ships in the convoy.

On July 10, the ominous sound of seventy German bombers was heard approaching the British coastline. Their destination was at first unknown, but it soon emerged their targets were the docks in the busy port cities of South Wales. This was the first blow in a German aerial assault that would come to be known as the Battle of Britain. Radar, Chain Home, Fighter Command, and the operational tactics and procedures developed since those early days at Biggin Hill, all were about to be put to the ultimate test. Two days later the bombers came again—this time targeting the ports of Aberdeen and Cardiff. It was a day that saw the first major successes for Britain's fighters, too, when Hurricanes launched from the East Anglian airfields behind the radar screen were able to destroy four German bombers attacking a convoy off the coast of Suffolk. Under the dark cloud of this newly opened assault on the mainland of Europe's last bastion of democracy, back in the United States the Democratic Party's convention opened its doors to delegates in Chicago.

The mood of the Republican Party convention in Philadelphia a few weeks before had been overwhelmingly in favor of nonintervention in the war. Wendell Willkie was selected to be the Republicans' presidential candidate by a margin of 654 to 318 over Senator Taft. It looked as though the American electorate was going to be faced with a genuine choice come November, because the day after Willkie's nomination, Roosevelt signed another Navy Bill providing for the construction of forty-five more ships and making no less than $550 million available to finance these and similar projects. The president was looking more and more willing to get involved.

On July 18, Roosevelt achieved the Democratic nomination for an unprecedented third term. The following day he delivered his acceptance speech, via radio from the White House, telling the delegates that he had asked himself whether he had the right, as commander-in-chief of the Army and Navy, to call on the men and women of America to serve their country while at the same time declining to serve their country himself. The world, he said, was now dominated by armed aggression "aimed at the form of government, the kind of society that we in the United States have chosen and established for ourselves." He told them that all private plans, including his own, "[had] been in a sense repealed by an overriding public danger. In the face of that public danger all those who can be of service to the republic have no choice but to offer themselves for service in those capacities for which they may be fitted."[12]

Across the Atlantic the task of matching men to the jobs for which they were "fitted" was proving difficult. On June 30, 1940, just nine days after the fateful meeting which precipitated Tizard's resignation, Patrick Blackett, ever a loyal supporter of Sir Henry, followed suit, sending a letter to Archibald Sinclair, the Secretary of State for Air, resigning from the CSSAD. "I am in entire agreement with Sir Henry Tizard," he wrote, "that the Committee cannot function usefully under the present conditions." He was particularly keen to stress the fact that Britain could ill afford to lose the talents of a man like Tizard at a moment of such dire peril, arguing strongly for Tizard to be given "a position of high executive authority in relation to the development and operational trials of new weapons."[13] It was a call that would soon be answered.

Tizard had not been the first choice to head the mission. The second half of July had seen a fevered round of consultations in which various eminently qualified people were offered the job, including Air Marshal Joubert, and even Frederick Lindemann. All of them had turned it down on the grounds that they were already involved in vital war work. The task of finding someone to lead it had been given to Lord Beaverbrook, the newspaper magnate recruited by Churchill to head the Ministry of Aircraft Production. He had been brought into government principally because of his legendary dynamism and powers of persuasion. But now, just two months into the job, he seemed to be having difficulty filling this vital post. Some who seemed to have a good claim had been ruled out. Hill was thought, as a physiologist, to be the wrong man to engage in complex discussions with American radar experts. Watson-Watt, who clearly thought himself to be absolutely the right man for the job, was judged to be insufficiently diplomatic. Cockcroft was considered, but his war work at the Ministry of Supply meant that his superiors were unwilling to spare him for more than a few weeks. Tizard, by contrast, was out of a job and available. He was a scientist and an airman. He had been at the center of the development of radar. He was a capable administrator and a good diplomat. He would turn out to be the perfect choice.

One of those who clearly recognized Tizard's qualities was Air Marshal A. W. Tedder. In a letter sent on July 16, 1940, in support of Tizard he wrote, "it is clear that by far the best leader would be Sir Henry Tizard, who combines the wide scientific knowledge needed with the knowledge of operational requirements, and who has an international standing as a scientist."[14] Despite his advocacy of an exchange, Tizard himself seems not to have thought very much of the plans at first. His diary records that: "I said I would not go unless I was given a free hand and that I did not want to go in any case unless the mission was really important for the war. It looked to me at first sight as a rather neat method of getting a rather troublesome person out of the way for a time!"[15]

Telephone calls were placed to the service departments to find out who they had in mind to attach to the mission. It was early days, true, but Tizard discovered the Air Ministry was thinking of sending only a flight lieutenant, and the War Office no one at all. Tizard made his feelings very clear. He would only take on the task if the military agreed to attach officers of seniority and front-line experience to the mission.

A meeting of the War Cabinet on July 25 saw them throw their considerable weight behind the mission. They instructed all service branches to prepare lists of the technology to be handed over to the Americans, making it clear that the disclosure must be full and open. Two days later Churchill himself made it known that he wanted the mission to go ahead as soon as possible. Tizard now had approval from the very top, and the service departments had begun to fall into line. The way seemed clear for rapid progress. But he was still to encounter a few as yet unforeseen hurdles.

In the Battle of Britain not everything was going Britain's way. Chain Home might be aiding the fighters to pick up raiding bombers crossing the English coastline, but out in the North Sea and the English Channel the Luftwaffe was beginning to take control. Following a series of heavy losses of destroyers, on July 28, the Navy was forced to withdraw all its remaining destroyers from Dover to Portsmouth. They could simply not afford any further losses and the withdrawal was a tacit admission of the ability of the Luftwaffe to control the Channel Narrows during the hours of daylight. To mount the expected invasion the Germans would have to establish both air superiority and control of the Channel. They were now coming dangerously close to the second of these objectives. Once more Churchill decided to plead with Roosevelt to send his World War I destroyers. The terms of Churchill's cable of July 31 were candid, almost intimate. It is as if he was writing to the ally that the United States had not yet become.

> Former Naval Person to President Roosevelt
> 31.vii.40
>
> It is some time since I ventured to cable personally to you, and many things, both good and bad, have happened in between. It has now become most urgent for you to let us have the destroyers, motor-boats and flying-boats for which we have asked. The Germans have the whole French coastline from which to launch U-boats and dive-bomber attacks upon our trade and food . . . We have a large construction of destroyers and anti-U-boat craft coming forward, but the next three or four months open the gap of which I have previously told you. Latterly the air attack on our shipping has become injurious. In the last ten days we have had the following destroyers sunk: *Brazen*, *Codrington*, *Delight*, *Wren*, and the following damaged: *Beagle*, *Boreas*, *Brilliant*, *Griffin*, *Montrose*, *Walpole*, *Whitshed*; total, eleven. All this in advance of any attempt which may be made at invasion! Destroyers are frightfully vulnerable to air bombing, and yet they must be held in the air-bombing area to prevent seaborne invasion. We could not sustain the present rate of casualties for long, and if we cannot get a substantial reinforcement the whole

> fate of the war may be decided by this minor and easily-remediable factor . . . I am beginning to feel very hopeful about this war if we can get round the next three or four months. The air is holding well. We are hitting that man hard, both in repelling attacks and in bombing Germany. But the loss of destroyers by air attack may well be so serious as to break down our defence of the food and trade routes across the Atlantic.[16]

Roosevelt was sympathetic to the request for destroyers, and now, at last, he seemed to be getting some help from influential sectors of public opinion. William Allen White's Committee to Defend America by Aiding the Allies (CDAAA) and a new organization, the Century Group, founded on the day Italy entered the war, had both been pressing hard for some time for the transfer of the veteran destroyers to the British. Churchill's telegram had, in fact, followed advice from Lothian that this might be an opportune moment to set Britain's case out to the president. But both Lothian and Churchill knew that even now some really substantial military quid pro quo would have to be offered in order to overcome the increasingly vocal objections of isolationists in the United States.

A little over a year earlier, in June 1939, King George VI and Queen Elizabeth had traveled to America for a royal visit. The aim was to bolster support from across the Atlantic for Britain in any future war. One of the less formal moments of the visit had been an invitation to the Roosevelts' country home at Hyde Park, during which the president talked frankly with the king about America's attitude toward the developing situation in Europe. He first brought up the suggestion of a neutrality patrol to protect the Atlantic approaches to the Western hemisphere—a patrol which would offer joint protection to both America and Britain's possessions in the region. In tandem with that suggestion he floated the idea of Britain granting the United States the right to establish military bases in Bermuda, Newfoundland, and Britain's Caribbean islands, something the U.S. Navy regarded as highly desirable. More strident voices were suggesting the bases might be handed over in return for the cancellation of World War I debts, but in the summer of 1939 this would have been a step too far for both governments.

A year on, the idea of the grant of base rights in return for the destroyers began to emerge as a serious possibility. The difficulty was, just as Roosevelt needed to be seen to be trading the destroyers for something tangible, so Churchill did not want to be seen to be giving away sovereignty over Britain's colonial possessions in exchange for fifty out-of-date warships. Lothian had therefore, in May, suggested that his government "should seriously consider officially a formal offer to U.S. to allow it to construct aerodromes and naval stations in British Islands which are important to its security," hoping that this would nip in the bud the pressures for these bases to be fully ceded in exchange for debt. Churchill had not been keen, but the world had turned a great deal since May, and now, in the darker situation of July, Churchill and his cabinet agreed that Lothian should offer base rights in return for the destroyers.

On August 1, Tizard's appointment received the imprimatur of the prime minister when he met with Churchill at 10 Downing Street. Churchill wanted him to know that he believed Tizard to be the man for the job. Tizard wanted an assurance that he would be given a free hand. Churchill concurred, and asked Tizard to list his requirements. Apart from stating categorically that the mission should include a high-ranking officer from each of the three services, Tizard had one further stipulation: he wanted to be able to fill two posts personally, with a curious pairing of John Cockcroft, one of the world's leading nuclear physicists, and a "junior radar scientist." Clearly he already had in mind Eddie Bowen, then just twenty-nine and a graduate of Swansea University College, rather than Cambridge. Things were going well, and Tizard left Downing Street confident that he and the prime minister were singing from the same hymn sheet. Within a few short hours that confidence was to be undone.

This meeting between the two men had taken place late in the afternoon. Churchill was unwell, and being looked after by his wife, yet there had been no sign of his being distracted, or unsure about the situation. But just a few hours later the prime minister's position had changed. Later that same evening Tizard telephoned the Ministry of Aircraft Production only to discover that Churchill had withdrawn his authorization for the mission to go to America. Only the use of the word "temporarily" prevented the situation looking completely disastrous.

The difficulty lay in the fact that the two men had quite different perspectives on the goals of the mission. Tizard's genius had been to recognize that this was not the time for Britain to repeat the mistakes of the past. The idea of exchanging technological secrets had been tried before, several times. But always that "fetish of secrecy," on both sides of the Atlantic, had left such exchanges stillborn. Tizard's experience on the CSSAD and his deep understanding of the vital role that could be (indeed he was certain would be) played by electronic aids in the air war, had led him to propose a mission to *give away* Britain's secrets. This was what made it so revolutionary, and so hard to stomach for military men and politicians conditioned to regard secrets as just that—secrets. Tizard had long ago come to the opinion that Britain's beleaguered industry simply would not be able to develop and manufacture this equipment at the necessary speed and scale. Only by accessing the enormous resources, in both development and manufacturing, of an America that was as yet not embroiled in war, could Britain's forces be supplied with the new weapons they so desperately needed. Only by giving away those new technologies could such resources be accessed. Anything that could be obtained in exchange would be a bonus. Churchill, in contrast, still clung to the idea that these secrets were something to be bargained with. If not directly for American secrets, then perhaps in some other way. They were a trump card, for sure, but they could be played only once. Churchill had clearly decided that now was not the time.

As August began Churchill had other things on his mind. On August 1, no response had yet been received to his desperate cabled plea of the previous day, but that evening, in Washington, Lord Lothian met with Navy Secretary, Frank Knox. It was a meeting at which Lothian did all he could to convey Britain's plight, so much so that, when Knox spoke to his cabinet colleague Harold Ickes on the telephone the following morning, he reported that Lothian had been "almost tearful in his pleas for help and help quickly." Knox told Ickes that he had been sympathetic. "However," he had said, "we can't do anything without legislation and we can't hope to get a bill through Congress without showing that we have received adequate consideration from England."[17] Knox had then put forward the idea of a direct swap of destroyers for bases. Lothian replied that though he was personally sympathetic to the idea, he could not make any commitment on the part of his government. Knox went on to tell Ickes that he planned to bring up the matter at the cabinet meeting due to take place later that day, August 2, and at this cabinet meeting the decision was taken to press ahead with the destroyers for bases deal, as Roosevelt himself recorded:

> At Cabinet meeting, in afternoon, long discussion in regard to devising ways and means to sell directly or indirectly fifty or sixty World War old destroyers to Great Britain. It was the general opinion, without any dissenting voice, that the survival of the British Isles under German attack might very possibly depend on their getting these destroyers.
>
> It was agreed that legislation to accomplish this was necessary.
>
> It was agreed that such legislation if asked for by me without any preliminaries would meet with defeat or interminable delay in reaching a vote.
>
> It was agreed that the British be approached through Lord Lothian to find out if they would agree to give positive assurances that the British Navy, in the event of German success in Great Britain, would not under any conceivable circumstances fall into the hands of the Germans and that if such assurances could be received and made public, the opposition in the Congress would be greatly lessened. I suggested that we try to get further assurance from the British that the ships of their Navy would not be sunk but would sail for North America or British Empire ports where they would remain afloat and available.[18]

Churchill's cable on the last day of July had been the first for several weeks, weeks in which Britain's position had been becoming ever more perilous. The cable was, indeed, a distress call, but on August 1 it must have seemed that not only was that plea not going to be instantly met, but that the Americans were seeking to extract a heavy price in return for "50 or 60 of your old destroyers." Churchill, for his part of the bargain, was being asked to give two things that would greatly endanger the fragile morale of the British people—ceding rights in some of its colonial possessions, and once again to guarantee that the Royal Navy would not be allowed to fall into German hands in the event of a defeat—something that Churchill felt he dared not even countenance in public. Small wonder, then, that in 10 Downing Street on the evening of August 1, Britain's entire set of technological secrets should have been regarded as a bargaining chip to be withheld.

Nevertheless in Washington after that crucial cabinet meeting Roosevelt contacted Lothian personally to inform him of the outcome. The necessary legislation for the destroyers exchange was to be put before Congress, and Roosevelt was doing all he could to ensure it was not opposed by the Republicans. Roosevelt was keenly aware both of the dangers in this course of action, and the necessity of taking it. Offering material support to the last standing democracy in Europe could easily make the United States an enemy where currently it had none. It was a decision that had to be made on a national basis.

The two countries also held different opinions about the underlying purpose of the proposed scientific mission, and about the relative value of the technologies on offer. To the British, radar was the key. Over the past five years money and effort had been poured into the practical application of radio direction finding. The result was a unique defensive chain of coastal radar stations that even now were showing their worth as Britain's first line of defense against attacking bombers. For the British, radar was far and away the most important secret they had to offer. They were firmly convinced that this was a technology unmatched by anything in American laboratories, let alone working out there on the front line and delivering practical results. Radar, surely, was what the mission was all about. The Americans, though, took a somewhat different view. Radar was important, yes. So much so that they already had systems in development. But America was not as vulnerable to the bomber as Britain. Neither was it at war, and so Americans did not place such a premium on defensive radar as did the British. The Americans had a broader agenda, and once the mission had been agreed to, both the Army and Navy began to put together comprehensive lists of the information and equipment they wanted, alongside that which they would be prepared, eventually, to disclose.

On August 2, the Americans' lists arrived in London. The Navy wanted information not just on radar, but also on radio, sonar, power-driven gun turrets, and the latest developments in armor plating and self-sealing fuel tanks. They wanted to see designs for the newest aircraft engines, propellers, and aerodynamics. And they wanted to know what the British had learned about mines and minesweeping in their operations at sea. To this the Army added tanks, armored cars, antitank guns, explosives, aircraft armament, balloon barrages, and chemical weapons. The Americans, it seemed, wanted the lot. Yet despite the presidential green light, the American service departments had still not yet formally agreed to hand over anything at all. One thing, though, had already been decided—details of the Norden bombsight were not to be disclosed.

Much work remained to be done then in sorting out the details of the mission, and now that it had been agreed at the very highest level, notwithstanding the "temporary" stay, there was no time to lose. The first two weeks of August saw Tizard gradually assembling a small team able to cope with the broad range of American interests while holding the British focus on radar. Things were coming together fast, and now it seemed that only Churchill stood in the way of the mission's quickly setting sail for America. On August 5, when the Foreign Secretary Lord Halifax heard of the delay, he wrote immediately to Churchill reminding him of the significance Roosevelt placed on the early arrival of the mission. Halifax copied his letter to the rest of the War Cabinet, convinced that they would help pressure Churchill into authorizing the mission's departure. Next day the military weighed in alongside the Foreign Office when General Hastings Ismay wrote Churchill in similar terms. When Churchill came to office Ismay was his personal choice to be his chief military assistant and staff officer. In that capacity Ismay served as the principal link between Churchill and the Chiefs of Staff Committee and his views carried a lot of weight in Downing Street. By the end of those two days Churchill would have been in no doubt of the strength of feeling of those around him who were convinced that the mission should be held up no longer. On August 6, he duly gave way and replied to Halifax authorizing the mission to proceed. "I am anxious," he said, "that they should proceed as soon as possible and under the most favourable auspices. The check of August 1st was only temporary."[19]

Four days later, on August 10, what had now become known as the British Technical Mission to the United States was able to hold its first meeting at Thames House by the riverside on London's Millbank. Sir Henry Tizard opened proceedings by explaining that the cabinet had invited him to lead a mission to exchange secret technical information with the U.S. government. The mission, he said, had full authority to disclose all secret information in the possession of His Majesty's Government. During the month that had passed since Lord Lothian's proposal landed on Roosevelt's desk in the Oval Office, Tizard had already assembled around him the six key men who were to form the core of his mission. Perhaps the most distinguished of them was John Cockcroft, who would later describe the mission as having been "magnificently organised" by Tizard: "One of his great inspirations had been to bring a mixed team of serving officers and scientists [together]. So our American friends for the first time heard civilians holding forth with authority on instruments of war with their service friends following on with practical experiences."[20]

It was undoubtedly a stroke of genius, though the choice of just six individuals to carry this heavy burden would be vital. The chosen six were: Brigadier F. C. Wallace, a distinguished Army officer who had just returned from Dunkirk, where he had been in charge of the anti-aircraft defenses; Captain H. W. Faulkner, recently returned from naval duty off the coast of Norway; Group Captain F. L. Pearce, of the RAF's Coastal Command, who had led the first bombing raids on the German battleship *Scharnhorst* as it lay in a Norwegian harbor; Cockcroft himself, a distinguished Cambridge physicist then attached to the Ministry of Supply; Arthur Woodward-Nutt, from the Air Ministry, who would act as the Mission's Secretary; and Edward Bowen, recruited by Tizard to the vital role of radar specialist.

Bowen later recalled the stress that Tizard placed on the importance of building good working relationships between mission members and their American counterparts: "He left us in no doubt about the importance of the Mission, and the seriousness with which it was regarded by the Government. He also showed his wry sense of humour. Without diminishing the magnitude of the venture in any way, he hinted that between negotiating sessions in the U.S. there would be some equally serious socialising to do. Looking at us in his whimsical way, he indicated that it was the job of the military members of the Mission to shoulder that particular burden. Would the civilian members please exercise restraint and keep their heads clear!"[21]

Tizard had negotiated carefully with Churchill before accepting the job, and had insisted that the mission should be as open as possible, but it was clear that at this early stage the British were also considering holding back some items of their own. The minutes of that first meeting of the British Technical Mission record that: "The Chairman [Tizard] emphasised that it was quite likely that we should give more information than we received, but the object of the mission was largely to create goodwill. In general, he proposed to pass on full information about matters which had already reached a fairly advanced stage of development, but only very general indications of the lines on which we were working on subjects which were still in the very early stages."[22] It was clear that much would depend on the reception the Mission received when it arrived in Washington, and on the relationship the two teams would strike up.

The commission's members began to bring in all the samples and reports with which they would demonstrate to the Americans their latest cutting-edge developments in all the technologies of modern warfare. For days now the team, as it came together, had been frantically assembling all the documentation, films, and exhibits they could find. As Tizard had no office of his own, the task of gathering everything fell to Cockcroft, who put anything that landed on his desk into a large black deed box, little more than a suitcase, which was kept under close guard at his office at the Department of Supply in Savoy House in London. Cockcroft records that these included the micropup valves used to power existing AI and ASV sets, and "all the reports within our sphere of knowledge, R.D.F, No 3. predictor, Gyro (type 6) gun sight, rockets, jet propulsion." Altogether the meeting at Thames House listed some twenty-one different technologies, even including the latest developments in chemical warfare and jet propulsion. Samples were to be taken where possible, along with drawings and plans. In recent years films had become a vital part of the development process and had also been used for training purposes, particularly in the field of radar. These too were to be taken and disclosed. Even now, though, some restrictions were agreed. Jet engines were to be discussed only in general detail, and the Royal Navy insisted on holding back details of its latest magnetic mines.

The jewel in the mission's crown was the "precious magnetron," as Cockcroft called it. Working with Bowen on their contract, the GEC laboratories at Wembley had been given the task, or perhaps the opportunity, to build a working radar system using the cavity magnetron. In that hectic final week of July, just as the technical mission was being given the green light, Eric Megaw and his team at GEC were demonstrating just how much of a leap forward the cavity magnetron was. It was more powerful than their previously state-of-the-art 25 centimeter micropup by a wide margin, while at the same time being a fraction of the size and physically far more robust. The cavity magnetron, it was becoming plain to those precious few in the know, was the future of electronic warfare. But it was by no means clear at this point that the cavity magnetron would be handed over. That would depend on how open the Americans were going to be.

Cavity Magnetron No. 12, then, was by far the most important item in the box—the newest of all the secrets it contained. Since the beginning of the year Bowen had been liaising regularly with both EMI and GEC on their separate contracts with the Air Ministry to develop centimetric airborne radar. In the few weeks since it had been handed over, he had been closely following Megaw's development of Randall and Boot's experimental cavity magnetron into a prototype production model. Following a meeting of the liaison committee on August 5, Bowen took a subway train out to the GEC research laboratory at Wembley on the outskirts of London for a detailed briefing on how the production magnetron was built. It became plain to him at that meeting on August 7 that this new device would revolutionize not just his own field of airborne radar but many other aspects of warfare on which radar could be brought to bear.

Of the total of twelve magnetrons produced so far by Megaw's team, one had already been dispatched to AMRE (now renamed the Telecommunications Research Establishment) at Worth Matravers. Another had been sent to the Royal Navy's Signal School at Portsmouth. Megaw put the remaining ten through their paces on a test rig, and Bowen was allowed to choose the best of them to pass on to the Americans. His choice of No. 12 was to have what he subsequently described as an "interesting consequence." In fact for a few hours it would threaten the success of the entire mission. Arrangements were made for Bowen to return a few days later to collect the magnetron, along with the detailed accounts of its construction and blueprints giving the precise internal dimensions. On August 11, he returned to Wembley to pick up cavity magnetron No. 12. As he traveled back to London, no one sitting in the same carriage of the Metropolitan Line train would have had the remotest idea that the young Welshman was carrying a piece of electronic hardware on which Britain's very survival would depend.

All these preparations were taking place against a background of uncertainty, not just about whether the mission itself would go ahead, but also about the whole nature of the relationship between the United Kingdom and the United States. Churchill's prevarication about the mission and its role in fact mirrored doubts on the other side of the Atlantic, because the two-month period from the middle of June saw a fundamental reshaping of transatlantic international relations. The secret scientific and technical mission that Tizard and his men were about to begin would take place silently, in the shadows of the destroyers for bases deal; yet once the conclusion of that deal had opened the door to a new cooperation between the two English-speaking powers, the Tizard mission would do much more to turn it into a practical reality.

On August 13, Churchill sent a cable accepting the leasing of military bases on British colonial soil to the Americans for ninety-nine-years. The details had yet to be finalized, but the deal was effectively done. Roosevelt would be allowed to release "50 or 60 of your old destroyers" on the basis of a trade, thus avoiding the necessity to put the matter before the still isolationist Congress.[23] As if to underline the signatures yet to be applied to the contract, Lothian suggested that British destroyer crews should be immediately dispatched across the Atlantic in order to be ready and waiting as soon as the destroyers became available. It would, he said, help to impress the urgency of the case on Congress. Work was immediately put in hand to assemble the crews, and within three weeks they were embarking for Canada. When the liner *Duchess of Richmond* steamed slowly out of Liverpool on August 28, its passenger list included a thousand officers and men of the Royal Navy on their way to await the arrival of the American destroyers—and the handful of men who made up the top-secret Tizard mission.

ELEVEN

First Contact

THE PLAN WAS THAT TIZARD SHOULD FLY OUT AHEAD of the rest of the mission to prepare the ground for their arrival by sea with their box of tricks. Now that the mission was officially under way it became clear that it was not a moment too soon. *Adlertag* "Eagle's Day," Tuesday, August 13, had arrived. The German plan was a four-day assault to clear the sky over southern England of RAF fighter defenses to establish a platform for a September invasion. On this day the Luftwaffe's planes flew some 1,500 sorties. Despite being on the back foot and significantly outnumbered, the RAF managed to get fighters into the air on no fewer than seven hundred occasions. The bright August sky over southern England was filled with dogfights for hour upon hour. In the final tally the Germans lost forty-five planes, the British just thirteen. Chain Home was doing its job, but the Battle of Britain was escalating by the day.

The morning of August 14 dawned inauspiciously for the Luftwaffe. Weather conditions were not nearly as good as the previous day. The raids that day were much lighter, a welcome respite for the RAF, who were able to reequip some of their downed pilots. One man who did take to the air that day was Tizard himself, boarding the flying boat *Clare* in Poole harbor on England's south coast. The *Clare* took off bound for Montreal via Foynes in Ireland and Botwood in Newfoundland. RAF Group Captain Pearce traveled with him. Pearce recorded that the journey took 25 hours and 37 minutes.

As the two men traveled from Newfoundland southwest to Montreal, back in Europe the RAF, and the civilians below, were about to endure the most critical 24 hours of the Battle of Britain. On August 15, the Germans flew almost 1,800 sorties, the highest number in the entire battle. RAF fighters took to the air against them almost a thousand times. This time the attacks came not just over southern England, but across the north too. The Luftwaffe had miscalculated badly. They believed that Britain's fighter force was almost spent, and that all available planes had been diverted to the south in response to the earlier attacks. But Air Marshal Dowding had been using the advance information provided by the radar shield to ration his forces carefully, putting fighters into the air only sparingly, determined to have considerable resources up his sleeve when things got worse. In particular he had not called on those squadrons defending the north, and so, against all their expectations, the Luftwaffe's specially adapted Messerschmitt 110s, with extra fuel tanks replacing the rear-gunners for the long flight across the North Sea, found themselves hopelessly exposed. A force of Junkers 88s got through to inflict serious damage on an RAF airfield at Driffield in Yorkshire, but the rest of the raiders suffered heavily in the attack in the north, losing twenty-three aircraft from a force of about one hundred and fifty. They shot down not a single RAF aircraft. Further south, though, the story was different. By the end of the day the RAF had lost a total of thirty-four fighters and several RAF airfields were left badly damaged. Overall German losses, though, were far higher: seventy-five in all. Britain had survived its hardest day.

Three thousand miles away Tizard and Pearce traveled on from Montreal to the Canadian capital of Ottawa. Part of Tizard's brief, while he passed through Canada on the way to Washington, was to "establish relations with the Canadian authorities with the objective of commencing research and manufacture of required equipment in that country." When A. V. Hill visited Canada earlier in the year during his own mission, he had been impressed by the quality of the scientists he met. But until the dramatic escalation of war in the spring of 1940, little effort had been put into the scientific development of weapons in Canada. Hill clearly saw Canada as something of a missed opportunity, because his report made a powerful case for Canadian capabilities: "Everything I saw and heard convinced me that there had been a great lack of imagination and foresight on our part in failing to make full use of the excellent facilities and personnel available in Canada,"[1] and argued "Canada may have to exercise a fundamental role in Imperial Defence, and that we should do everything possible to get things going there to improve their resources."[2]

Tizard must have been disappointed, then, with what he found. He knew that the Canadians had been briefed by the British on the state of their radar development back in the spring of 1939; but without British technical assistance, and with little money having been made available by the Canadian government, all that had been achieved was a single coastal defense radar set, and that was primitive by comparison with what the British had achieved at home.

One man who could enlighten Tizard further was Ralph Fowler. A professor of mathematical physics at Trinity College, Cambridge, and the son-in-law of Sir Ernest Rutherford, he was one of that impressive array of Cambridge physicists who had been brought in to revolutionize the radar effort at an early stage. Shortly before the war started Fowler had been appointed to head Britain's National Physical Laboratory, a position he soon had to resign after suffering a stroke. He had, however, recovered well, and when the war began he made it clear that he would do whatever he could to assist the war effort. One of Hill's recommendations had been the establishment of the post of British scientific attaché in Ottawa in an effort to meld Canadian science into the war effort. Fowler, he thought, would be just the man. Fowler accepted and traveled over to Canada in July. So by the time of Tizard's arrival Fowler was fully informed about the state of Canada's technological development potential, and also of the caliber of its scientists.

Fowler's briefing to Tizard was less than encouraging. The Canadian scientists were indeed impressive, but the resources provided by their government were woefully inadequate. So far this unfortunate situation had been exacerbated by the lack of significant information being made available about developments in Britain.

The Canadians had, however, recently been sent a couple of early examples of gun-laying radar, and they were quick to see the superiority of the British equipment, deciding, despite the difficulties of having to adapt it to North American power supplies, that they should abandon their own design in favor of the British. On Fowler's advice, Tizard arranged for two of Canada's top men to join them in Washington as soon as they were able: C. J. McKenzie, the head of the National Research Council, and Air Vice Marshal A. V. Stedman. They were also to be accompanied by Colonel H. E. Tabor, the Master General of Canada's Ordnance Office. Their presence was to strengthen the mission, but also to lay the groundwork for scientific cooperation between Canada and the United States at a time when the Canadians themselves were reluctant to put money in. Now, at last, it was time for the main body of Tizard's work to begin.

A week after his arrival in North America Tizard and Group Captain Pearce boarded a train bound for Washington. Throughout that week the news from Britain had been heavy with accounts of fierce resistance to the German aerial onslaught and serious fighting in the Mediterranean theater. August 16 and 18 both saw major Luftwaffe raids; but again it seemed the RAF had managed to beat them back. German losses over the two days totaled almost 130 aircraft, while the British had seen fifty fighters shot down. The RAF's Bomber Command had struck back, more for the effect on morale than with the expectation of success. A night raid on armaments factories at Leuna, Germany, had achieved little but to expose the virtual impossibility of accurate bombing at night; but there had been moderately successful raids against Fiat factories in Turin and the Caproni aero-plant in Milan.

The principal business of the week, though, was the almost overwhelming Luftwaffe assault on southern England, with German planes crossing the coastline more than fifteen hundred times in a single day. Yet the RAF's vastly outnumbered Fighter Command was resolute, and on August 20 Churchill made another powerful speech, a paean of praise to the young RAF fighter pilots who were holding the world's finest air force at bay: "Never in the field of human conflict was so much owed by so many to so few."[3] On this day, too, came the official British announcement that bases would be leased to the

United States. The "destroyers for bases" deal had finally been sealed. It was surely a good omen for the success of the mission.

Political agreement for the exchange of secrets to go ahead had fallen into place on August 13, just before Tizard left Great Britain, though the Americans still had to decide exactly what they were prepared to hand over. If equipment and designs were to be handed over to the British, there was the small matter of patent rights to consider. The technologies shortly to be discussed were to be developed and produced by commercial concerns that would not take kindly to giving away their trade secrets. Even at the beginning of August, American policy was that no information could be handed over that was not directly under the control of the U.S. government. It was soon recognized, though, that this was completely unworkable, because just about all the weapons systems being developed included components patented by private companies. The policy would have to be changed. By the time Tizard arrived in Washington on August 22, it had been decided that information could be given on any military technology not *specifically* excluded. In all cases where private trade rights were involved the British would be required to secure the rights to manufacture in return for suitable financial compensation.

That evening the battle on the British home front entered another dimension when German heavy artillery, now stationed on the French coast, began to bombard the town of Dover, a key port just twenty miles or so across the Channel in England. The British returned fire from their fourteen-inch coastal battery. It would turn out to be the first of many cross-Channel artillery duels.

When Tizard finally arrived in Washington his first port of call was the British Embassy for a meeting with the ambassador. Lord Lothian had, of course, been one of the most active proponents of the mission, but Tizard was annoyed to discover that he had done little to prepare for their arrival. No offices had been set aside, and neither had any clerical support been made available. What is more, the embassy was short of space and had been unable to find room for Tizard and Pearce, let alone for the full mission, who would be arriving in a little over a fortnight. Fortunately Tizard, ever the diligent administrator, had already arranged with the NRC in Canada for the assistance of the two of their secretaries, and within a couple of days had established his headquarters in rooms at the nearby Shoreham Hotel overlooking Rock Creek Park and just a ten-minute drive from Capitol Hill. Now, with an office and the nucleus of the team around him, the work started in earnest. He had just two weeks to build a rapport with the Americans so that the mission could begin its work in an atmosphere of trust.

On his very first evening in Washington Tizard dined at the home of Admiral Potts, the British naval attaché, with Admiral W. S. Anderson, the U.S. Navy's director of naval intelligence and the man designated responsible for liaison with the Tizard mission. This was followed by a meeting the next day with Anderson and Frank Knox, the Secretary of the Navy. It was the opening gambit in setting the ground rules for the exchange. His second evening in the capital was spent in the company of Colonel J. A. Crane, in charge of foreign liaison for the U.S. Army's General Staff.

Then, somewhat incongruously it must have seemed to Tizard, came the weekend. He had left behind a world threatened with imminent disaster, where every day, every hour, counted. Each minute that ticked by seemed to take Britain closer to the final reckoning. But Washington was still at peace, and the weekend was still the weekend. Odd though he must have found it, Sir Henry was no doubt glad of a little rest and respite. Over the previous two days the German attacks on southern England had showed no sign of abating. But now the bombers were coming with ever-stronger fighter support. Their targets were shifting, too, to cities all across the United Kingdom and, in particular, to ports. And they came at night, more than 150 at a time, stretched out across the country. On August 24, as Tizard dressed for dinner in his comfortable Washington hotel room, the night bombers targeting airfields around London dropped their bombs on the suburb of Croydon after failing to find their target—the nearby RAF airfield at Kenley. This was in contravention of strict orders, but it provoked a backlash when Churchill sent RAF bombers to attack Berlin the following night. The damage was slight but it was another turn of the screw.

As Monday, August 26, dawned in Washington, the pace of Tizard's rounds moved quickly into top gear. The day began with the first formal meeting with Admiral Anderson, following their social meetings of the previous week. Anderson was quick to say that right from the start the Americans planned to discuss the vast majority of their own secrets openly, with the specific exclusion of the Norden bombsight. It was immediately clear to Tizard that the exchange was getting off on the right foot, because such American openness had been anything but guaranteed when he left England. In fact he was now far less interested in the Norden sight than the Americans may have thought. What Tizard wanted was U.S. manufacturing capacity above all else. Anderson then set before Tizard the list of technical information the U.S. Navy was looking for, and the two men began to divide the items into groups, some of which would be discussed in general meetings, while others would be delegated to special committees.

That afternoon Tizard made the ten-minute drive south toward the White House. He was accompanied by Lothian, but because of the extreme secrecy surrounding the mission they made their way in through a back door in order to avoid the assembled newsreel cameras and reporters standing guard at the main entrance. The significance of the meeting lay more in the fact that it was taking place than in what Roosevelt had to say. The president wanted to demonstrate to Tizard that the exchange had the presidential stamp of approval. The details were to be handled by the scientists and the military. Tizard was clearly impressed by the president, whom he described as "a most attractive personality."[4] Roosevelt, for his part, talked about the need to prepare the United States for war. He spoke about the draft conscription bill he wanted to push through Congress. Confident that it would succeed, he nevertheless worried that this effort might well cost him the election in November. Not that it mattered, he said, provided that it meant the United States was better prepared if war came.

The president went on to explain that the Norden bombsight held a totemic place in America's defensive arsenal. It would be political suicide to hand it over so that a single crashed airplane could put it in the hands of a potential future enemy even before American airmen had been called upon to use it. Nevertheless, if the British demonstrated that the Germans already had a similar technology, the position might change. Tizard asked that the dimensions of the sight should be made available in order that British bombers could be modified to take it, in case it later became part of the anticipated exchange. He updated the president on the current state of British radar development and when Roosevelt heard that films demonstrating the techniques involved would shortly be arriving, he expressed an interest in seeing the films for himself. As Tizard knew, this was all grist to the mill, and by the time Tizard and Lothian slipped back out of the White House, the cordial tone of the meeting must have done much to convince them of American goodwill. Sir Henry could now prepare confidently for another crucial meeting the following morning, this time at the War Department with Henry Stimson, the new Secretary of War.

This would be a vital encounter. Stimson was impressed by Tizard, "a very nice and sensible British scientist."[5] In the months to come Stimson would prove to be a crucial link between scientists building on the work of the mission and their political masters in Washington. He urged Tizard to meet as soon as possible with Vannevar Bush at the newly constituted National Defense Research Committee. On radar itself he suggested two names, Karl Compton and Alfred Loomis. A founding committee member of the NDRC, Compton was one of the most powerful scientific administrators in United States and a close ally of Bush's. Helping to run the Paris office of America's National Research Council back in 1918, Compton was also one of the few Americans who had experienced at first hand the benefits of mutual scientific exchange. By 1930 he had become president of the Massachusetts Institute of Technology, and in the years that followed had done much to advance the relationship between engineering and science. He and Tizard—a scientist, skilled administrator, and former Rector of Imperial College, Britain's leading engineering school—found they had much in common. Stimson's other suggestion, Alfred Lee Loomis, a millionaire financier and gifted amateur physicist who happened to be Stimson's cousin, was to play a key role in the creative explosion that resulted from Tizard's mission.

Alfred Loomis was fifty-two years old and already had a glittering career behind him. Having trained as an attorney, he spent World War I at the Army's Aberdeen Proving Ground, where his work on ballistics showed him to be a talented scientist. During his time there he invented the Aberdeen chronograph, the first portable instrument for measuring the muzzle velocity and force of bullets. After the war he turned away from both army and law and went into the banking business with his brother-in-law, Landon Thorne. During the 1920s electric power was beginning to spread rapidly across rural America. Loomis was quick to spot the potential. After buying into the long-established banking house of Bonbright and Company, Loomis and Thorne engineered a takeover. Rescuing it from the brink of bankruptcy, they quickly turned Bonbright into a new style of investment bank, specializing in the emerging public utilities. Within a few short years Loomis and his brother-in-law had become the bankers to the electrification of America. The rewards were colossal, so much so that they were able to engage in that most expensive of rich man's hobbies, the America's Cup. When they found time to relax, they would retire to their private 17,000-acre estate on Hilton Head Island, South Carolina, for riding, fishing, and hunting.

Many of those who grew rich in the 1920s, of course, lost much of their wealth in the crash of 1929. Not Alfred Loomis. Convinced that the seemingly endless rise in stock prices must be coming to an end, Loomis managed to convert his substantial wealth into cash before the bubble burst. He was one of the few Wall Street financiers to survive the crash virtually unscathed. When the stock market bottomed out he and Thorne were among a select few who had capital to invest. In the aftermath Loomis was able to buy cheap and multiply his considerable fortune many times over. Now Loomis and his brother-in-law were able to run their hugely expensive J-Class America's Cup yachts on their own, at a time when even the more famously wealthy, like the Vanderbilts, were operating theirs in syndicates. But Loomis had bigger and better plans for what to do with his money. His ambitions were on a higher plane than most bankers.

In the 1930s Alfred Loomis began to put much of his fortune to good use to advance science research, supporting some of the most cutting-edge projects of the day. But he was much more than merely a philanthropist of science. He was himself by now a scientist of some distinction. A dilettante (in the true and positive sense of the word), he built himself a state-of-the-art laboratory in Tuxedo Park, the exclusive country estate where he lived some forty miles up the Hudson River from New York City. Here he was able to attract some of America's brightest young scientific talents, along with many of the most prestigious names in science, some of whom had already found themselves exiled from the darkening situation in Europe. In the decade of the Depression, and with the shadow of fascism spreading slowly across Europe, Loomis played host to a string of scientific luminaries. Niels Bohr, Werner Heisenberg, Enrico Fermi, and even Albert Einstein were guests at Loomis's laboratory in the Tower House. Here too he was able to indulge his own passions. One of those passions was radar.

Politicians were not the only people worried about the threat that the bomber represented to America's historical sense of invulnerability. Vannevar Bush was acutely aware of the situation and the premium it placed on the development of an early warning system that could see the bombers coming. So far, the Army had put virtually no resources into developing defensive radar, but the U.S. Navy had a better record. They had a working design, but little funding to develop it further. The previous year Bush had written to former president Hoover: "The whole world situation would be much altered if there was an effective defense against bombing by aircraft. There are promising devices, not now being developed to my knowledge, which warrant intense effort. This would be true even if the promise of success were small, and I believe it is certainly not negligible."[6]

Bush was placing great store on the work that would be done by Division D of the new NDRC, the division working on radar, which was to be headed by Karl Compton. Compton in turn would place great store on Loomis, especially once it came to meetings with Tizard's men, because Loomis was not only well versed in the latest radar developments at home, but he had also already visited England and met some of its top scientists on their home ground. The preceding month, in the weeks before the Battle of Britain had begun in earnest, Loomis had traveled across the Atlantic. That he was able to meet and speak with leading British radar scientists owed much, of course, to his hospitality at Tuxedo Park and his Tower House research laboratories.[7] His reputation had gone before him. What he discovered from scientists already dealing with the realities of war was their urgent need for a microwave system delivering precision guidance for night fighters and anti-aircraft guns. But Loomis knew, from his own research, just how difficult this was.

Back at Tuxedo Park he had, in early summer 1940, assembled a team of young scientists who were hard at work trying to develop some sort of portable microwave detection system. He was well aware of the problem of generating enough power, having experimented with the only two devices that then seemed to hold out any hope of producing sufficient power—the klystron and something called the Sloan–Marshall resnatron that had been developed at Berkeley under the auspices of his close friend the renowned physicist Ernest Lawrence. A pioneer nuclear physicist, Lawrence was busy working on a project to build a 184-inch cyclotron in California. A key part of this work was the need to generate a high-speed electron stream. One potential method was using a Sloan tube to generate radio waves for radar. Lawrence reacted cautiously when the idea was suggested to him by Loomis, and responded: "It seems to me . . . that a matter of this sort is not susceptible to complete paper analysis, and I would be in favor of someone's experimentally developing the multiple resonator arrangement to find out its good and bad points. You know as well as I do that experimental work always brings to light things of which one does not think by any amount of cerebration."[8]

Loomis was not the sort of man to walk away from such a challenge. He was a great believer in experimentation and had spent much of his life, and a not inconsiderable portion of his wealth, facilitating it. Throughout the summer Loomis's small band of researchers had been working on the construction of an experimental portable radar system based on the Doppler principle that waves increase in frequency as an object moves toward an observer, and decrease as it moves away. They built a "target" consisting of a series of copper wires on a moving belt. When the belt was set up in a nearby wood they were able to use the laboratory set to determine whether the belt was moving or not. The next step was to see if it could detect something useful, like a moving vehicle. They would also try to make the device itself mobile. This second set was installed in a large delivery van that Loomis had bought to be converted into a mobile radar installation. Painted in the traditional Tuxedo Park colors of green with gold trim, and bearing the legend "Loomis Laboratories," it became something of a public manifestation of the otherwise discreet research work going on at Tower House. On one of its first outings the truck was driven a couple of miles south to the Tuxedo Park golf course, the nearest patch of land devoid of trees, where it was pointed at the automobiles on South Gate Road running through the golf course toward Tuxedo Lake. The on-board indicators clearly showed not only the approaching automobiles, but also the speed at which they were traveling. Inside the Loomis Laboratories van was the world's first radar speed gun. The results were astonishingly accurate and all those who took part were fully aware of the potential of the system they had created.

Soon they were able to detect motorboats on the nearby lake, and after driving down to Bendix Airport they had considerable success monitoring the movements of small planes out to a distance of some two miles. This mobile set used an 8.6 centimeter Klystron to generate its microwaves. It was a first, and it was microwave radar, but it was too weak and too cumbersome—the apparatus filled up most of a van—to offer a realistic military option. Nonetheless it was enough to convince Loomis of the potential of microwave radar, and of the desperate need for that elusive new power source that could generate microwaves at far greater energy. It was enough to convince Karl Compton and the NDRC too. Compton had the highest regard for Loomis, and microwave radar was right at the top of the wish list of D Section, his fiefdom in the NDRC. What is more, Loomis sat on the board of MIT and was a generous financial supporter; he was well known to James Conant and Vannevar Bush. Loomis was an obvious choice, then, to head up D-1, the microwave subcommittee Compton decided to set up to oversee the hunt for battle-ready microwave systems.

More or less the first thing that Compton and Loomis discovered was the walls of secrecy around the radar developments being undertaken by the two services. Their Congressional brief "to coordinate, supervise, and conduct scientific research on the problems underlying the development, production, and use of mechanisms and devices of warfare,"[9] it quickly became apparent that the Navy and Army were making progress, but that both were equally reluctant to share the advances they made. Compton and Loomis immediately saw the danger of their committee becoming a mere pipeline for communications between the services, and also that this was by no means enough. That they should share their information was a step in the right direction; that they should work together was the goal. The urbane socialite Loomis was just the man to bring the two sides together in a convivial atmosphere. Loomis decided that the third meeting of his Microwave Committee on July 30 would be preceded by a large dinner in the pleasant surroundings of the sumptuous suite he kept at Washington's Wardman Park Hotel. Bush, Compton, and Frank Jewett, head of the Bell Telephone's laboratories, were all invited, along with other members of the committee, and, of course, senior representatives of the Army's and Navy's radar development programs. Ostensibly this was an opportunity for the military men to meet members of the committee; in fact its prime function was for them to meet each other and to begin to establish an effective collaboration. The discussion went on long into the night, and did much to lay the groundwork for the future development of radar in the U.S. military.

The military has traditionally been notoriously conservative when it comes to the technical preparations for war. In the summer of 1940 the U.S. military was not at war, but it was already gearing up for it, and so while they might grudgingly accept Karl Compton, an academic thrust upon them by executive decree, they might be less willing to accept Alfred Loomis, a millionaire dilettante scientist whose subcommittee was apparently tasked with developing the weapons of the next war. Loomis would need all his considerable personal skills if he were to build effective working relationships. One immediate outcome of the meeting was that arrangements were made for Compton and Loomis to visit the services' various research establishments to see the work that was actually already in hand.

August's fact-finding tour culminated in an invitation to attend the army maneuvers set to take place at Ogdensburg in upstate New York. Compton and Loomis were invited to watch an early field test of the Army's newest secret radar system, the SCR-268. Compton was later to observe that the special treatment they were afforded in being given close access to scrutinize the operation of the SCR-268 made them "the envy of the high officers attending the maneuvers because not even the generals were allowed to get near enough the equipment to find out anything about its operation."[10] Their high standing was further underlined by their arrival in a private airplane belonging to Loomis's cousin—the Secretary of War, Henry Stimson. If the military doubted for one moment the ability of these scientists to get things done, they were now surely aware that these men had something of a hotline to the highest levels of political power.

The SCR-268 was a revelation to Loomis. The research work his team had been carrying out at Tuxedo Park operated on the Doppler principle, using continuous waves and monitoring the frequency change in the returning echo. Much of the research work that had been carried out by the U.S. military up to now had also used continuous wave transmission. But the SCR-268 was different—it was a pulsed radar system. This was the form of radar transmission that had been recommended by Robert Watson-Watt in the UK way back in 1935, and was the system that had been enthusiastically adopted in Britain for the Chain Home system that was now proving so effective. Under the threat of imminent war the British had taken up and developed a technique that had in fact first been put to use in America some ten years earlier, while the U.S. military had since mainly traveled another route. Now, though, with the SCR-268, the Army seemed finally to have produced an effective pulsed radar system. This technique for measuring the distance to the earth's ionosphere, first developed by Merle Tuve and Gregory Breit at the Carnegie Institute in 1925, did not transmit a continuous wave of radio energy, but a series of tiny pulses, each no more than a few microseconds in duration. After each pulse the system waited (in radio silence) for the faint returning echo. As the transmitted radio waves travel at the speed of light, these pulses can be transmitted many times in a single second. Thousands of such readings, showing up as a blip on a cathode ray tube, show the track of a moving target. As Compton and Loomis watched the Army's field test of the SCR-268 there can have been little doubt that this was the way of the future. If pulsed radar could be somehow combined with a powerful source of microwaves so as to miniaturize the equipment and sharpen the beam, then the power of this new weapon would be fully unlocked.

On the evening of August 28, Tizard sat down to a pleasant but pivotal dinner. He was being entertained by Vannevar Bush, head of the NDRC, at the Cosmos Club in Washington. A private social club, it had for some sixty years been at the center of the social life of Washington's intellectual and scientific mandarins. Its membership at the time was restricted to "men distinguished in science, literature and the arts." This was very much a meeting of minds between first-class scientists and consummate administrators. Tizard told Bush about the extraordinary dividends that were being reaped in the UK through the involvement of university scientists in the war effort. Bush listened attentively, then offered Tizard some advice of his own. Keen though he was that his scientists should talk directly with Tizard and his men, he thought it imperative that Tizard first built his own relationships with the military. The alternative would play into the hands of those who wanted to see the whole thing as a transatlantic conspiracy of scientists to wrest control away from the military.

Earlier that day Tizard's team had been considerably strengthened. Since July the Royal Navy already had a substantial technical presence in Canada and the United States. The British Admiralty Technical Mission (BATM) had offices in Ottawa and New York and technical experts stationed there whose job was to liaise with manufacturing industry on the construction and/or purchase of war material. These were the men, the Admiralty had decided, who were best placed to assist Tizard. On Wednesday, August 28, Vice Admiral Evans, who was heading up the BATM, met with Tizard and agreed to attach three of his technical experts to Tizard's mission. The addition of Commander Crossman, J. W. Drabble, and A. E. H. Pew to the team significantly strengthened their ability to talk authoritatively about the very latest developments in naval technology. Pew was in Canada to try to kick-start an ASDIC (sonar) manufacturing industry, while Crossman and Drabble were specialists in radio technology. These were men who could talk to American technical experts in their own language, and by the following day they were already deep in discussion with the Navy's Bureau of Ordnance and Ships. These talks produced instant results when it quickly became apparent that since World War I the navies of Britain and America had developed fundamentally different answers to the same problem in the use of sonar. For almost twenty years the two navies had grappled with the problem of interference to a ship's sonar readings caused by its own reflected waves. The Americans had attempted to solve the problem by developing more sophisticated transmitters, where the British concentrated on designing their sonar equipment so as to reduce the reflected noise. Each had had only limited success. But now, as technical experts from the two navies met to exchange information for the first time, they quickly realised that a combination of the two methods would produce significantly better results. In a single afternoon the potential benefits of the Technical Exchange had been powerfully demonstrated.

But there was more to come. The following day, these early members of the mission paid a visit to the Naval Research Laboratory (NRL) at Anacostia, where they were able to examine a device that would have a radical impact on the design of future radar systems. One particularly tricky technical difficulty with radar design was that the power generated by the transmitter was huge in relation to the faint returning echo from a small metallic object some tens of miles away. All British radar up to this point had therefore utilized separate transmitting and receiving antennas, the receiver having to be shielded from the high power generated by the transmitter. With space at a premium in ships and, particularly, aircraft, no way had yet been found to combine the two, despite the obvious space-saving advantages. During the visit to the NRL Crossman and Drabble were amazed to discover that this problem had been cracked some two years earlier by American scientists, who had developed a device called a duplexer. The duplexer was essentially an electronic switch that closed the receiver's circuit automatically during the transmitting phase of the cycle, in order that the highly sensitive receiver should not be overwhelmed by the power of the transmitted pulses. With the duplexer it became possible to use a single antenna both to transmit and to receive. The potential benefits for air interception radar, in particular, were plain to see, and Crossman and Drabble arranged for it to be examined as soon as possible by the mission's radar expert (and specialist in the unique problems of airborne equipment) Eddie Bowen as soon as he arrived in the country.

Tizard, meanwhile, was returning the compliment at the War Department, where he was giving the Americans their first detailed rundown of the Chain Home and Chain Home Low Systems, and also revealing the existence of a new system, which, when perfected, would become an integral part of radar control systems for the rest of the war. Before the Battle of Britain, RAF fighters had been fitted with a system codenamed "Pipsqueak," a clockwork mechanism which broadcast a special radio signal for fourteen seconds of every minute. The Fighter Command sector stations had receivers to monitor the "Pipsqueak" broadcasts so that the bearings of the signal could be telephoned to the sector control rooms, where they would be triangulated and the fighter's location plotted. In this way the RAF fighters were able to signal their position automatically. It was a crucial aid to the controllers, but crude. The far more sophisticated Identification of Friend or Foe system, or IFF, would be a radar-based version enabling radar operators to determine which of the signals on their screens were their own aircraft and which the enemy's. The Americans were quick to see its merits and, when fully developed, the common use of IFF by British and American pilots would be of vital importance once they had begun fighting as allies.

The frankness of these exchanges did much to cement in American minds not only the fact that these men arriving from Britain could be trusted, but also that they might have something to offer. Tizard's men too were beginning to realize that despite the British superiority in operational experience, American development of radar systems was, in many respects, significantly advanced and that the British themselves had much to learn from them. In fact Tizard's openness about IFF had so impressed General Mauborgne, the head of the U.S. Army's Signal Corps, that he invited Tizard to visit the Corps radar research center at Fort Monmouth, New Jersey. There Tizard was able to see for himself the SCR-268 gun-laying set that had been such a revelation to Loomis and Compton. The principle of reciprocity was taking shape, and there was still more than a week to go before the majority of the mission members and their precious cargo would arrive.

TWELVE

The *Duchess* of Richmond

As Sir Henry Tizard sat down to dinner with Vannevar Bush on the evening of August 28 in the civilized surroundings of the Cosmos Club, across the Atlantic Eddie Bowen was fast asleep in the Cumberland Hotel near London's Marble Arch. Beneath his bed was a case containing Britain's wartime secrets. Just two things had been held back—the precise details of the very latest developments in jet propulsion and information about some German magnetic mines that had recently been recovered. Nestling in among the plans and manuals was cavity magnetron No. 12. A prior arrangement had been made for the case to be kept overnight in the manager's safe. But when Bowen had checked in the previous evening. the case had turned out to be too big, and with an early start for the boat train to Liverpool, Bowen decided the safest place for it was under his bed.

In the preceding few days the task of assembling everything the six members of the mission would take with them had been completed, and now the moment had come for them to make their separate ways to Liverpool to rendezvous for their passage across the Atlantic. To Bowen had fallen the responsibility of ensuring the safe delivery of the most concentrated collection of wartime secrets that Britain had ever known. Bowen's journey to Euston station the following morning was almost as mundane as his trip on the London Underground with the cavity magnetron a couple of weeks earlier, as he recalled:

> For reasons best known to himself, the taxi driver refused to let me carry the box inside the taxi and insisted that he should strap it on the roof. There was no time to argue, so we made the short run to Euston with that supremely important piece of luggage prominently displayed on the roof. With my other luggage, the box was more than I could handle, so I called a porter and told him to head for the Liverpool train. He grabbed the box, put it on his shoulder and headed off so fast that (an old cross-country runner and still pretty fit) I had great difficulty keeping up with him. He got well ahead and the only way of keeping track of him was to watch the box weaving its way through the mass of heads in front. A first class seat had been reserved for me, but beyond that I did not know what to expect.[1]

Settling down into his reserved seat, Bowen was pleased to see that the entire compartment had in fact been reserved. The blinds on the corridor windows had been pulled shut and there were large signs bearing the word RESERVED on the outside of the compartment. Shortly before the train pulled out, though, a smartly dressed man opened the door, sat down opposite Bowen, and began reading a copy of the *Times*. Bowen was puzzled. The man clearly knew what he was doing, and Bowen judged it better to speak only when he was spoken to. A few moments later, things became a good deal clearer when the sliding door shot open and a couple of stragglers stepped in to take the vacant seats. The smartly dressed man ordered the two men to leave the compartment instantly. Bowen was impressed: "It was not what he said but how he said it . . . for the first time, I realised that the precious cargo was under some kind of protection."[2]

Bowen had been instructed to stay put when the train arrived in Liverpool. He was to wait for the arrival of an army detachment, who would take care of the box and its safe transport to the waiting Canadian Pacific steamship *Duchess of Richmond*. But as the train

pulled in there was no sign of an army presence on the platform. So he waited. And so did his traveling companion. They sat for a few minutes in silence until they heard the sound of marching feet. Looking out of the window Bowen saw a dozen fully armed soldiers coming along the platform. They collected the box and marched off in the direction of the quayside, leaving Bowen, who until now had had sole responsibility for his cargo, to make his own way to the ship. He arrived on board and checked with the captain that the box had been safely delivered and was being kept under guard on the bridge. The captain told him he had been given instructions on what to do in the event of an enemy attack. Such was the significance of the contents of this small black box that the captain had been told that if there was any danger of the ship being lost, it was to be dropped overboard and allowed to sink to the bottom of the ocean.

By the end of the afternoon the full complement of the ship's passengers—including more a thousand Royal Navy personnel, whose task of manning the American destroyers was still secret, and the members of the mission—Bowen, Wallace, Faulkner, Cockcroft, and Woodward-Nutt—had arrived on board, and the *Duchess of Richmond* began pulling slowly away from the quayside. A perilous journey across an ocean patroled by seemingly unstoppable U-boats lay ahead. Just the previous day the *Dunvegan Castle*, a 15,000-ton armed merchant cruiser, had been sunk by a U-boat torpedo to the west of Ireland. The safety of neutral America seemed far away, and the mission was to be in danger sooner than they might have imagined.

The summer sun was sinking slowly in the west as the liner gently eased away from its Liverpool moorings. A hundred or so miles to the east, a flight of German bombers was heading across the North Sea on a raid targeting Liverpool and those same docks. Gladstone dock had been built especially for the great transatlantic liners and was situated right at the mouth of the River Mersey, giving immediate access to the Irish Sea. On that night it was perhaps just as well, as no sooner was the *Duchess of Richmond* under way than the German raiders arrived over the city and the bombing began. Bowen recalled the evening: "We left our berth as darkness fell and headed for the Irish Sea. We had not gone very far before bombs started to fall around us. They were not particularly accurate, but were sufficient to stop the ship. There had been sporadic night bombing in London during the preceding few weeks, but no one took it seriously and this was presumably more of the same directed at Liverpool. We spent the night in the Roads and started off next morning."[3]

Bowen's bravado may have owed much to the proverbial British stiff upper lip, but the bombers now finally arriving over London were far more significant than he thought, and the raid he'd just witnessed on Liverpool was the first of four consecutive nights when the bombers targeted the city. That they were virtually unhindered was stark testimony to the helplessness of anti-aircraft artillery and the near impossibility of air interception by fighters at night. Without radar (and microwave radar at that) little could be done against the night bomber coming singly and without pattern. Above their very heads that night was the principal reason why the mission's success was so vital.

The delay was only temporary, the danger less than had been feared, and by early the following morning the *Duchess* was under way with an escort of two destroyers. Once into open water it was able to build speed and the destroyers peeled away. Just twelve years old, it had been one of the more modern liners on the North Atlantic run when it was requisitioned for duty as a troop ship just two weeks earlier. At a little over 20,000 tons and with a top speed of nearly twenty knots, it was fast enough to outrun the German U-boats and all but the fastest of surface vessels. Like all other requisitioned liners, its speed was its defense, and it would make the crossing unaccompanied. An additional defensive measure, however, was a constant changing of course every twenty or thirty minutes so as to prevent the U-boats lying in wait. It would add many hours to the crossing and earn the ship the nickname "The Drunken Duchess" as it weaved its way across the Atlantic.

It must have seemed an age to Tizard's men, who were desperate to get to America to begin their work. As they listened to the BBC's radio broadcasts from the UK the news of increasingly heavy air raids only sharpened the sense of urgency. Nonetheless, there was nothing to be done and the eight-day crossing dragged on. From time to time a question would arise among the team that required consulting documents locked in the box tucked away safely on the bridge.

Woodward-Nutt, the mission secretary, was the only person allowed access to the box on its journey, and after it had been extracted several times he commented on how heavy and difficult it was to shift. It had been assumed that in the event of the *Duchess* being sunk by enemy fire the box, too, would go straight to the bottom. But would it? To a brilliant physicist like Cockcroft this was something of a challenge and, with nothing better to do, he sat down one evening in the bar before dinner to work the answer out. His conclusion was that the assumption was false, and that there was every chance that the box would bob about on the surface waiting to be picked up by the enemy. Something would have to be done.

Rumors quickly spread on board ship that one of the world's leading scientists was indeed among the passengers and Cockcroft was soon asked to alleviate the troops' boredom with a lecture. He was happy to do so, but what could he talk about? Here he was, sitting on top of the very latest developments in military technology, every one of which would have fascinated the troops—but of course he could say nothing about any of them. Eventually he decided to play it safe. Eddie Bowen was in the audience: "Cockcroft gave it some thought and decided to tell them about atomic energy. He gave a marvellous lecture in very elementary terms about radioactivity and the energy residing in the nucleus. He invented a new unit called the battleship-foot: the amount of energy necessary to lift a battleship one foot out of the water and, following a line of argument advanced by Rutherford, described how a single cup of water had enough energy to lift a battleship many feet into the air and probably break its back. This, he explained, was a very safe subject to talk about because there was no hope at all that such a thing would be achieved during the present war!"[4]

The reality of the present war had continued unabated during their passage, and Britain's defenses were on the point of cracking under the strain. The German tactic of trying to neutralize the RAF on the ground was meeting with considerable success. Concentrated attacks on RAF airfields had severely limited the capacity of many and put some completely out of action, including Biggin Hill, Manston, West Malling, Lympne, and Hawkinge. Debden and Hornchurch had been badly damaged. The RAF was now losing aircraft at almost the same rate as the Luftwaffe. And the supply of replacement aircraft was now in peril too, with the Luftwaffe scoring hits on several aircraft factories. More or less the only good news was the formal announcement from Washington on September 2 of the deal to make the old American destroyers available to the Royal Navy.

Eventually the interminable voyage came to an end and the *Duchess* steamed into harbor at Halifax, Nova Scotia. On the quayside was an American armed guard waiting to spirit the box and all its secrets away to Washington. As it was being loaded into a heavily armored vehicle, Bowen noticed that a neat pattern of holes had been drilled in each end so that it would quickly fill with water once it were thrown overboard. Sometimes the simplest solutions are the best.

There was one other revelation for the mission at Halifax. As the *Duchess* edged into harbor another ship was in the process of docking. It was a four-stack American destroyer, the first of the "fifty or sixty of your oldest destroyers" which Churchill had so painstakingly negotiated with Roosevelt. Now it became clear why the *Duchess* had been transporting those thousand naval ratings and officers across the Atlantic. Not that the aged ships would really be missed. Admirals Stark and Furlong had had to clench their teeth in the face of considerable opposition, to certify the destroyers as surplus to U.S. requirements. It was a politically dangerous maneuver, but already, by September 9, it had become truer than might have been envisaged. On that day in Washington a new $5.5 billion appropriations bill was passed into law. Contracts were to be placed for 210 vessels for the navy, among them seven battleships and twelve aircraft carriers. Those ancient destroyers were soon to be genuinely surplus after all.

Now the mission divided its forces. While Wallace, Faulkner, and Woodward-Nutt headed directly to Washington to meet up with Tizard, Pearce, Bowen and Cockcroft had work to do in Canada. Although the black case that Bowen had so carefully transported from London to Liverpool contained all Britain's paper secrets, the cavity magnetron was unique in being the only piece of equipment small enough to be carried in this way. There were, however, other pieces of hardware Tizard had arranged to be shipped across the

Atlantic so they could be demonstrated by members of the mission. Principal among these was Bowen's own prize exhibit, a working example of his MkI ASV system, a working airborne air to surface vessel radar set operating on a metric wavelength. The plan was to fit it into a Royal Canadian Air Force Hudson, which would then be flown down to Washington to demonstrate the progress Britain had already made on airborne radar. Sadly, and somewhat ominously, this powerfully persuasive plan fell at the first hurdle. All the way across the Atlantic Bowen had assumed that the several crates addressed to the National Research Council and stowed safely on the bridge of the *Duchess* contained his precious ASV set. Now, however, it became clear that this was not the case. The crates were indeed destined for the NRC but had nothing to do with the Tizard mission and contained a completely different consignment of goods. Someone in London, it appeared, had taken the decision that instead of sending MkI equipment it would be better to wait until the new MkII system had been fully tried and tested and was in production. It would be a long wait, and once they were in Washington the mixup was a considerable source of embarrassment.

Notwithstanding this setback, Bowen and Cockcroft traveled to Ottawa to finalize arrangements with the NRC for Canadian liaison with the exchange. The remainder of the party arrived in Washington on Sunday, September 8, while Bowen flew down to join them three days later. Out in the North Atlantic the U-boat menace continued to grow when, on the night the *Duchess* arrived in Halifax, U-47 attacked convoy SC-2 sinking three merchant vessels. By attacking at night and on the surface the U-boats could neutralize the sonar defenses of the convoy. Clearly what was needed was exactly the sort of equipment that had been left behind in London.

But London had other things on its mind. On Saturday, September 7, German tactics fundamentally changed. Incensed by the RAF's raid on Berlin the previous week Hitler had decided to put London to the sword. Suddenly the bombers were switched from their task of putting the airfields out of action, in which they had all but succeeded. It was a major break for the RAF, who were now able to regroup. But for the people of London it meant the nightmare of the Blitz had begun. Hitler and Goering had committed the entire

2nd Air Fleet to the daytime assault—some 500 bombers with 600 fighters in support. The first wave of attacks in the afternoon brought 300 bombers and the full 600 fighters, almost a thousand planes, droning over East London to attack targets in the London docks. This complete and sudden switch of tactics took British defenses by surprise. By late afternoon much of eastern London was in flames. Even the arrival of dusk did nothing to halt the onslaught. At night another 250 bombers arrived, and for once the problem of target-finding evaporated as London was illuminated by the still-burning fires from the afternoon's offensive. This first day of the Blitz was a stark illustration of the danger of concerted night bombing. There was almost nothing the RAF could do to repel the raiders in the absence of a fully functioning system of radar guidance for night fighters. The first few prototypes were coming through, but, for the moment, the problem had arrived well in advance of the solution.

THIRTEEN

Unparalleled Power

THE BLITZ AT HOME WAS PROMINENT IN THEIR MINDS when the members of the British Scientific and Technical Mission to the United States sat down together in Washington's Shoreham Hotel on the morning of September 9, 1940. This was their first full meeting on American soil, the first of the internal briefing meetings that would happen every day the mission was at work. On this opening day Tizard divided the team up into eight specialist groups reflecting both the material they had brought with them and the areas of American interest: radar operational methods; technical details of radar; anti-aircraft gunnery, fire control, directors, and searchlights; aircraft armament, turrets, and tracers; defense against low-flying aircraft; proximity and influence fuses; aircraft and aircraft engines; and means of distinguishing friendly from enemy aircraft and ships.

He told the men that they were free to discuss any technical matters whatsoever with members of the U.S. services, but (in anticipation of some tricky discussions still to come on the issue of patents and property rights), that they were not to have technical discussions with any nonservice personnel unless specifically authorized by

Tizard himself. Then he issued another carefully thought-out instruction: Faulkner, Pearce, and Wallace, the three serving officers on his team, "should take every opportunity that naturally presents itself to discuss war experience with officers of the U.S. forces."[1] Bowen, the youngest of the party, had not yet arrived from Ottawa, otherwise he would surely have remembered Tizard's speech to them on that first day in London, when he warned that there was "some . . . serious socialising to do."[2]

The mission on which they now set out unfolded over the next three months in distinct phases. The first ten days were taken up primarily with the British sharing with the Americans all the information they had brought with them across the Atlantic, together with the crucial lessons they had learned from both combat experience and captured German technology. The next phase consisted largely of a series of reverse revelations, this time with the Americans opening the doors on all their secret development projects, with the exception, of course, of the Norden bombsight. But as they sat down to their opening meetings there were still some considerable obstacles to be negotiated on the road ahead.

That first week was notable for the different perspectives of the two sides of this exchange. To the British the mission was all about radar, but on the American side the initial focus of the military was less on radar than on securing access to a broad range of technologies. Radar they saw primarily as a defense against bombers, bombers that were not (yet) capable of crossing the great oceans that protected America. But the Americans, too, were in for an enlightening week or so, as they realized the huge strides the British had made in building radar out of its passive, defensive base into a weapon that had enormous potential in many fields of conflict. The difference, according to Eddie Bowen, Tizard's radar specialist on the mission, lay in the fact that in the previous five years American radar "had existed in only ones and twos and had seen little service use."[3] In contrast, Britain had invested £1 million (equivalent to over $435 million today)[4] at the end of 1935 and had built a vast network of operational radar stations defending the entire eastern coastline.

During that first week a vast array of technologies were discussed in detail and a bewildering amount of information was exchanged.

The British Army supplied details of their anti-tank weapons and information they had gleaned about German tanks. Information on explosives was exchanged, and techniques for controlling artillery were discussed in detail. Manuals and drawings of all British naval anti-aircraft guns were supplied, and visits were arranged for the U.S. Navy personnel to see most of them on ships in harbor at Halifax. Anti-aircraft gunnery was one thing on which the British had been gaining an uncomfortable amount of experience in recent weeks.

The sight and sound of an anti-aircraft battery blazing away at the raiders overhead may be good for the spirits of those on the ground, but hitting a bomber travelling at over 200 mph some two or three miles above the ground was exceedingly difficult. Apart from the time taken for a shell to climb to ten or fifteen thousand feet, and the fact that the target was moving at speed and could easily change course at any time, some method had to be used to calculate how far the plane would have flown by the time the shell arrived. In other words the aiming point had to be predicted, and here lay the key to improving the chances of successful anti-aircraft fire. The British provided details of their newest 3.7-inch gun and the quick-firing 40mm Bofors gun, a Swedish design that was proving very effective against aircraft. And with them came the Kerrison predictor, a fully automated mechanical anti-aircraft gunnery control system that was reputed to be the best of its type in the world. The Americans were quick to acknowledge that the Bofors gun and Kerrison predictor combined to make a powerful weapon, and this combination was soon adopted by both Army and Navy as their main medium anti-aircraft weapon system.

The Americans were also hugely impressed with briefings on the latest research into "proximity fuses," a possible breakthrough in anti-aircraft gunnery. The problem was that even if your predictor enabled you to get a shell where the target plane was about to arrive, a direct hit was almost impossible. Delayed-action timed fuses were a start, but with the crossing speeds of both shell and target being so high, the detonation could not be timed perfectly. Proximity fuses, however, seemed to offer a real solution. They were designed to self-detonate when they detected the presence of the target in close proximity, so that even a shell in the right area now became potentially deadly. It was a technology that would have startling benefits if it could be made to work properly. It was a technology that the Americans took up eagerly.

When it came to the issue of aircraft armament and bomber defenses, the Americans were keen to know about the recent development in Britain of power-driven turrets. The Americans had nothing quite like it and soon full production drawings, two sample turrets, and a mechanic were on their way across the Atlantic from the UK.

Aircraft engines were another key area of American interest, and the mission shared with them details of the latest internal combustion engines, particularly those powering the new generation of fighters. But they also had with them the first description of the prototype turbojet engine developed by Frank Whittle at Power Jets Ltd. Jet propulsion had been included in a list of "war winners" drawn up by the Air Ministry and the new Ministry of Aircraft Production early in 1940, and Whittle's engine was scheduled to be the power plant for a new Gloster fighter. That project, though, had been stalled by the desperate events of May, and the finer details of Whittle's revolutionary engine were not transported across the Atlantic. Nevertheless, American appetites were whetted, and gaining intelligence about this new power plant went to the top of the list once the U.S. Army Air Corps began making technical visits to Britain in 1941. Whittle's technology would subsequently become the basis for all American jet engine development.

Not all the technologies discussed were as successful or effective as the jet engine, however. Cockcroft also brought with him samples and drawings of barrage balloons, kites, and Lindemann's obsession, which had done such damage to Tizard, the aerial mine. He showed the Americans plans for the use of "unrotated projectiles" (UPs), as a defense against low-flying aircraft. We would now recognize these UPs as rockets, and though they didn't feature in quite the way envisaged in September 1940 they would play a significant role later in the war, and in particular on D-Day. Perhaps fortunately, another issue discussed that week never came to fruition; included on the list was information on chemical warfare, provided by chemists drafted in from the National Research Council of Canada in discussion with their American counterparts.

Tizard's genius in putting together the strangely varied membership of the mission, blending world-renowned scientists with young technicians and serving soldiers, now came into its own. Each presentation the British team made during that first week consisted of an unusual but powerful meld of two distinct voices, the civilian scientist, who would speak in depth on the technical details, and a serving officer, who could speak from personal experience about the device in combat. These officers gave separate talks on operational experience directly to their American counterparts. On many occasions they were able to bring these talks alive with training films, or even films or photographs taken in combat situations.

To the Americans, this combat footage was a sobering revelation. It brought home starkly the sacrifices that were being made in Europe—even as the news poured in of yet more unfolding agony. September 11 saw a daylight raid over London that got through to the target in force. The RAF almost came off worst, losing twenty-five aircraft to the Germans' twenty-nine. Buckingham Palace itself was hit by a bomb, though none of the royal family was hurt. Then came the huge offensive on September 15. It came in two waves, morning and afternoon. In the morning two hundred bombers with four hundred fighters in support took off for England. This time, though, unlike the previous weekend, the RAF was prepared, and fighters were scrambled as soon as the first radar traces appeared. The battle was very heavy, with RAF fighters picking the bomber fleet up as it crossed the coast and harrying it all the way to London. Over home territory, each fighter could exhaust its fuel and ammunition, return to base to refuel and rearm then take to the skies again to pick the raiders up on their return journey. The German bombers found themselves under constant attack for the whole of their time over British soil. A second wave in the afternoon brought similarly intense fighting, but this time the bombing was more accurate. In the final tally for this terrible day German bomber losses were severe indeed; overall the Germans had lost sixty aircraft to the RAF's twenty-six. The Luftwaffe's nose had been severely bloodied, and a change of tactics had been forced on the Germans. Large tracts of London lay smoldering, but the big daylight raids on the city were over.

Radar had done its job. Now it would be the turn of the night fighters, but they were still in dire need of radar help.

The pivotal position on Tizard's team was occupied by its most junior member. Eddie Bowen was still not yet thirty, but there were few people who could tell "Taffy" Bowen anything about the problems and potential of airborne radar. Now the full fruits of American research in the field were being laid before him. From what he saw it was clear that the cavity magnetron would astound them. It stood to revolutionize the American approach to radar.

Ten days into the mission proper, however, their most prized possession had still not been revealed. A. V. Hill had hinted at a new source of power during his visit earlier in the year, and Tizard had made oblique reference to an extraordinary advance that had been made in Britain during one of several informal meetings with Vannevar Bush that followed their dinner at the Cosmos Club. But it had been kept under wraps until the U.S. Navy finally lifted all its remaining restrictions on the exchange so that Bush and the NDRC—the organization that held the key to the manufacturing capacity Tizard had come in search of—were able to engage in full and frank discussions. Now it seemed, the moment for complete openness had arrived.

But for all the potential of the cavity magnetron, at this juncture it was nothing more than potential. The mission had nothing to show except the magnetron itself. Its invention had come too recently for anything like a working system to be developed, tested, and transported across the Atlantic. In contrast, metric-wave systems (whatever their limitations) were already up and running. Microwave-powered radar still seemed a long way off, and there was that strong feeling in America that microwaves were something "for the next war."

Nonetheless a far-sighted decision had been taken at the NDRC when they created D-1, a subcommittee to deal specifically with microwaves and put in charge of it a man with visionary abilities and the capacity to get things done—Alfred Loomis. Tizard knew right from the start that Loomis was a man who fully recognized the possibilities of microwave radar and the need to drive the research forward without delay. So it was to Loomis that he decided the magnetron should be disclosed.

Bowen and Cockcroft had not yet met Loomis when they received an invitation to evening drinks at his Wardman Park Hotel apartment. It was just a couple of hundred yards up Connecticut Avenue from their office suite at the Shoreham, but it was like entering a different world. The opulence of this millionaire's town residence was a complete contrast to the wartorn world they had left behind just a few days earlier. The evening was convivial from the start. In addition to Loomis himself, the small party included Karl Compton, Carrol Wilson (Bush's personal assistant), and Admiral Bowen of the Naval Research Laboratories. The admiral must have seemed like something of a cuckoo in the nest, because the Navy's agreement to meetings between Tizard's men and the NDRC had had a key stipulation. Talks could only take place on the understanding that a naval officer should be present at all discussions in which the department had an interest, and Admiral Bowen was an avowed skeptic about the exchange.

They settled down to what Loomis intended should be an agreeable but workmanlike meeting. Cockcroft later recalled "the rather doubtful opening with the U.S. officers suspicious as to whether we were putting all our cards on the table."[5] They can have had little idea how right they were. The Americans were aware that the British had something up their sleeve. During the previous week's meetings with the Army and Navy representatives the existence of the magnetron had been (in Eddie Bowen's words) "hinted at, but not described."[6] Now Loomis, ever the skillful negotiator, began the proceedings with a comprehensive and candid account of his Microwave Committee's recent tour of the industrial laboratories and research institutions where American work on microwaves was taking place. What the British learned that evening would provide them with a road map for the meetings and visits they themselves would request in the weeks to come. They discovered that the Americans had already made significant advances in transmission and receiving equipment at wavelengths as short as 40 to 50 centimeter. At the magic but elusive 10 centimeter wavelength Bell Telephone Labs and

General Electric had made impressive progress in developing receiver technology. Important work was also being carried out at MIT and Stanford. It was all very exciting, but there was clearly a gaping hole when it came to the ability to generate anything like sufficient power to transmit at the 10-centimeter wavelength.

It was a hole that could be plugged immediately by the cavity magnetron, and it was at this point that, Bowen says, "we quietly produced the magnetron and those present were shaken to learn that it could produce a full 10 kilowatts of pulsed power at a wavelength of 10 centimetres."[7] How shaken they were can be judged from the fact that that was roughly one thousand times the power that any American labs had yet been able to generate with their much larger Klystron tubes.[8] The Americans in the room were astounded. Loomis, in particular, was excited by the vistas it opened up and at the prospect that now lay enticingly before his Microwave Committee. With access to this unparalleled power source all manner of new devices now became possible. The challenge was going to lie in prioritizing the work, balancing America's future needs with the weapons the British needed right now to keep the Germans at bay.

Come the weekend that followed, the Washington newspapers carried a stark reminder of the desperate need for naval radar powerful enough to pick out U-boats operating on the surface as they shadowed convoys. In the North Atlantic, convoy HX-72 had been successfully hunted by a U-boat group that attacked constantly over a three-day period. Twelve ships had been sunk, seven of them during one night by a U-boat completely undetected by the convoy's escorts. The menace of the U-boat was growing ever more potent, and if it went unchecked Britain could be starved into submission.

The next week saw a welter of activity. Whatever reservations the Americans had seemed to have been swept away. For Cockcroft, the revelation of the magnetron had changed everything: "The disclosure was the key point and from then on we had no difficulties."[9] There was a trip to the USAAF headquarters at Langley Field. Bowen went to the CAA to discuss blind landing systems, while Cockcroft was busy comparing American anti-aircraft gunnery with British ideas about rocket defenses. Captain Faulkner flew to

Honolulu for an agreeable few days watching the U.S. battle fleet maneuvers in the Pacific. There were meetings on armor plating and a session where films were shown covering everything from training procedures to air combat. Bowen crossed the river to the Naval Research base at Anacostia; then went with Cockcroft on the first of many visits to Bell Telephone Laboratories. Next, it was a two-day trip with Cockcroft to Signal Corps's radar research center at Fort Monmouth, then back to Anacostia on September 27.

And on that same day Tizard, having just returned from a lightning trip to Ottawa, sat down for private discussions with Bush and Compton of the NDRC to agree on the future of radar research in the United States. Their minds were focused by events in Berlin, where Italy's foreign minister and Japan's ambassador were meeting with Hitler. They had come to sign the Tripartite Pact, promising that each would declare war on any country that joined the war against any one of them. Tizard heard that the Americans had made, in the week following the appearance of the magnetron, a key strategic decision. The Army and Navy would get on with producing battle-ready versions of the meter-wave kit they already had in development, but research on microwave radar was to be centralized in the hands of the NDRC's Microwave Committee. From now on the development of microwave radar would be masterminded by scientists, under the inspired leadership of Alfred Lee Loomis, and with the cavity magnetron at its heart.

And that work got under way in style when, on September 28, Cockcroft and Bowen flew into La Guardia airport in New York City. A limousine was waiting for them, with one of Loomis's drivers ready to whisk them the fifty miles or so northward to Tuxedo Park. It was a spectacular drive at a spectacular time of the year, and Bowen was entranced by the "beauty of the autumn colors in the elms, oaks and maples of the Catskills." Loomis greeted the two Englishmen as they arrived at the beautiful colonial mansion that was his summer home. Then they moved swiftly on for a tour of the place Cockcroft and Bowen were most anxious to see—the Tower House laboratory. Their final stop was an extraordinary new home Loomis had built near the highest point of Tuxedo Park—the Glass House, a building that must have been among the most modern anywhere in the world, with (then almost unheard-of in private homes) air-conditioning and "automatic everything else, much of it his own design" according to Bowen.[10]

There were two other guests at the Tower House that day—Ed Bowles, the professor of electrical engineering at MIT, and Carrol Wilson, Bush's personal assistant. First the Englishmen were given a demonstration of Loomis's Doppler radar, which was able to detect a small aircraft at a distance of two miles. Bowen, the radar veteran, was mildly impressed: "It would not have won the Battle of Britain but it showed the way their minds were working." The power unit was a Sperry Gyroscope Company klystron, and it's possible he might have thought more highly of it but for the cavity magnetron he had with him. The after-dinner conversation was all about centimetric systems and the improvements they could bring to the military's existing radar. The evening was convivial, the subject matter of their discussion desperately serious.

The following morning several others joined the party. Among them Loomis had summoned Hugh Willis, director of research at Sperry Gyroscope and a member of his Microwave Committee. Loomis had something that might interest him. The discussion continued throughout the day, and then in the evening Bowen and Cockcroft produced the magnetron. Again, the effect was formidable. Bowen described the "electric" atmosphere, as the potential of a device about the size of a man's hand began to sink in. It was quickly agreed that the Microwave Committee should offer to sponsor the development of the magnetron, and Loomis undertook to arrange for the Bell Telephone Laboratories to begin manufacture. The deal was not yet done, because neither Bell nor Tizard had agreed to it. But Loomis was used to getting what he wanted.

When Bowen returned to Washington to seek formal approval of the arrangements made at Tuxedo Park Tizard was, as he expected, enthusiastic and encouraged Bowen and Cockcroft to "push vigorously ahead."[11] Loomis meanwhile had already spoken to the Bell Telephone Laboratory to tell them that the Microwave Committee proposed to give them a contract for the production of a small batch of cavity magnetrons—always provided, of course, that the magnetron could deliver what was claimed for it. Bowen had no doubts on this score, though he was soon to find out that cavity magnetron No. 12 had a sting in its tail.

The morning of October 2, Sir Henry Tizard boarded a flight at Washington bound for La Guardia in New York. From there he would fly back to London via Bermuda. Now that the channels of communication were flowing freely his part in the mission that came to bear his name was over. But his legacy was only just beginning. He was leaving John Cockcroft in charge of the mission's ongoing projects in the United States. With the news from Britain it seemed just possible that there might be a glimmer of light at the end of the tunnel. One of Tizard's last acts before leaving the U.S. was to hand over a five-pound note to Cockcroft. It was to settle a bet—Tizard had staked a fiver that the Germans would have invaded Britain before he was ready to return.

Tizard asked Bowen to accompany him on that first leg of the flight. There was some information he had to pass on, and some advice on the way forward. He reaffirmed his delight that the Microwave Committee wanted to put the magnetron straight into production. Tizard knew how valuable this was, and urged Bowen to "push it for all you are worth."[12] Apart from anything else this would do much to correct the sense that the British could not deliver, a suspicion that had dogged the mission since the nonarrival of the MkI ASV set. Tizard had laid great emphasis, in his early days in Washington, on the fact that the mission would arrive with working airborne radar. So far they had not done so. Small wonder, then, if there were lingering doubts about the performance of the magnetron. He had however arranged for the shipment not just of the MkII ASV, but also a MkIV AI (airborne interception) set, and the latest identification of friend or foe , or IFF, system. It was absolutely vital, he said, that the systems should be made to perform properly once they arrived and he made it clear that Bowen should stay in the U.S. until the Americans had seen them in action. At La Guardia their ways separated. Tizard returned to an uncertain future in

Britain, while his words set Bowen on a path that would keep him on the other side of the Atlantic for many years to come.

To the Americans Bowen was a vital component of the microwave future. Loomis was keen to involve him from the very start. He had wasted no time in sharing the good news of what his Microwave Committee might soon be able to achieve. On the Monday after the momentous events at Tuxedo Park he returned to New York to bring the rest of the committee members up to speed. Then it was on to Washington for a meeting with his cousin, the Secretary of War. Stimson recorded that: "Alfred came in in the afternoon, full of excitement over his interviews with the British and with the scientists, and he was full of the benefits that we were getting out of the frank disclosure by the British of their inventions and discoveries of methods they have made since the war."[13]

Stimson was impressed, and called in General George C. Marshall to hear for himself what Loomis had to say. That evening Loomis joined a dinner party at Stimson's mansion home, Woodley, along with Brigadier Charles Lindemann, the British scientific attaché. The rest of the party was made up of Stimson's inner circle of key advisers and aides. In a telling vignette of the new political world Roosevelt was creating they were, to a man, Republicans working for a Democratic president.

In New York, the men of the Bell Telephone Company's laboratories were keen to get their hands on the magnetron. Following Loomis's phone call on Sunday, and his hurriedly arranged meeting with the rest of the Microwave Committee the following day, the Committee had placed an order with Bell to produce their own copy of the magnetron as soon as possible. Cavity magnetron No. 12 was back on its travels. This time Bowen was on his way with Cockcroft to Bell Labs's corporate headquarters on the edge of Greenwich Village in New York to meet first with their most senior executives, and then with the star members of Bell's radio tube research group. Bowen produced the magnetron itself and spread the blueprints and plans out before them. It quickly became clear that to manufacture it would not be a problem, but the Americans remained skeptical that it could do what was claimed. Bowen explained what they'd need—a powerful magnetic field, and a pulse anode able to generate 10,000 volts to power it. For this they'd have to ship it to the laboratories at Whippany across the river in New Jersey. The Bell men knew, from Bowen's descriptions, that they were looking at something that could potentially outstrip by a factor of seven even the most powerful device they had in development, and that it would, if it worked, be a thousand times more powerful than the most sophisticated device currently in use in America. Small wonder, then, that they were eager to fire it up as soon as possible.

Bowen, however, had pressing engagements for the next couple of days as he was travelling up to Boston for meetings at MIT. Bell would have to wait. On October 5, Bowen returned from Massachusetts to New Jersey in preparation for powering up the magnetron for the first time since it had left the GEC laboratory in Wembley some two months earlier. He spent a pleasant evening at the home of Bell's head of radio and television research, Ralph Bown, where he stayed the night. If he was anxious about whether the magnetron would perform the following morning, he didn't show it.

Next morning Bowen, Bown, Arthur Samuel, J. O. McNally and a handful of Bell's other scientists working on their radio research were gathered around the magnetron. This time, though, instead of being a mere object of wonderment, it was hooked up to a large electromagnet and waiting to burst into life. On these next few moments the military planning of two great nations would turn. Failure would have been a grave setback. Once the necessary magnetic field had been switched on and the magnetron had been connected to the modulator it was time to switch on the anode potential and see what happened. Bowen, understandably anxious, did so "very gingerly," but he need not have worried—almost immediately the machine began to produce a one-inch-long "glow discharge" from its output terminal. The magnetron had come up trumps.

The power generated was in fact in excess of what had been predicted, at 15 kilowatts, and the wavelength, the critical wavelength, was measured at precisely 9.8 centimeters. The Bell men were clearly stunned, but no more so than Bowen when he discovered that the Microwave Committee had already instructed Bell to build a batch of thirty magnetrons as soon as they had satisfied themselves that it worked. It had just passed that test with flying colors, and soon, it seemed, the magnetron would be in action. Not, though, for the first time.

Two other magnetrons, in addition to No. 12, had already been released by GEC—one to the Royal Navy and one to Bowen's former employers, the Air Ministry Research Establishment, newly renamed Telecommunications Research Establishment (TRE). At TRE they had already put the magnetron to work. In fact, just as Bowen was leaving GEC's Wembley laboratories with his valuable consignment, so TRE's magnetron was being prepared for its first field test the following day. In the wake of the emergence of the magnetron TRE had created a special Centimetre Group under Philip Dee, one of the brightest of the intake from Cambridge University's Cavendish Laboratory. Along with Alan Hodgkin and Bernard Lovell, who would be knighted after the war for his work on radio astronomy, Dee had built a portable microwave radar, mounted it on a swivel, and on August 12, just off the south coast of England near TRE's new home at Swanage, they had been able to detect a single aircraft traveling along the coast line. It was the first use of cavity magnetron microwave radar to track an aircraft—the harbinger of much to come.

The next day the equipment had performed perfectly when TRE put it to an even sterner test. This time Reg Batt, one of the team's assistants, had been asked to ride a bicycle along in front of a nearby cliff face carrying a sheet of metal. It was a crude simulation of a low-flying aircraft, but no existing radar set would have had any chance of detecting Batt on his bicycle, because of the overwhelming signals being returned from the ground below the beam. But, with the precision that microwave radar now suddenly offered, "a strong echo appeared on the cathode ray tube."[14] Just as its American counterpart the NDRC would a month or so later, TRE now threw itself wholeheartedly into microwave development. Dee's group was joined by others working on improving the receivers and antennae, or increasing the power output of the magnetron, and crucial contacts were made with groups working on the new technology in the Army and the Royal Navy, contacts that would feed into the famous, free-ranging brainstorming sessions at Worth Matravers that did much to ensure that the fighting men got the technology they needed rather than what someone thought they ought to have.

On October 7, Eddie Bowen took a telephone call from Mervin Kelly, Bell's head of research. Could he please come to New York—as quickly as possible. Something had gone awry. The following morning Bowen was on the first airplane out of Washington. Once in Bell's West Street headquarters, he was taken straight to a top-floor conference room where several Bell executives and scientists were waiting for him. The atmosphere was frosty. The magnetron, the famous No. 12, was placed on the table. Then the plans that Bowen and Cockcroft had brought across the Atlantic were laid out alongside it. Next to that, an X-ray photograph, clearly an image of the internal structure of the magnetron itself. Bowen leaned across the table to study the X-ray—and then sat back in amazement. What he had seen left him completely baffled. Looking again, from X-ray to plan, and back to X-ray, he checked once more what he had just seen with his own eyes. The blueprints he had carried across the Atlantic clearly showed six small circular cavities within the metal core of the magnetron, but the X-ray of the physical object revealed eight. Several stern faces stared blankly across the table at him. They were expecting an explanation. The trouble was, Bowen, Britain's leading radar man, had none.

Perhaps, Bowen suggested, he could send a cable to GEC back in London, where Eric Megaw might be able to offer some explanation. But the Americans were looking for an immediate clarification, and they were, after all, in the headquarters of the United States's leading telephone company. So, by what was then the "miracle" of transatlantic telephony, within a few short minutes Bowen was talking directly to Megaw. Bowen hoped that Megaw could account for the discrepancy the Americans had found. If he couldn't, then it looked as though all the initial goodwill and trust might be badly damaged. American production of microwave airborne interception sets was hanging in the balance.

At first Megaw couldn't think what had happened; then it came to him in a flash. Back in August, when Bowen traveled out to Wembley to collect a magnetron for the mission, there had been a total of twelve so far constructed. Two had already left GEC, for TRE and the Navy, and Bowen had been given the opportunity to pick which of the others he wanted to take. Not unreasonably, he'd chosen the latest to be built—number twelve of twelve. But no one at the time, or indeed until now, had realized the implication of that choice. In fact the first ten magnetrons had been built to the original specification (as set out in the drawings Bowen had taken to America) of six resonant cavities inside the block. The last two, though, had been treated as even more experimental. Number 11 had been built with seven cavities, and number 12 with the eight that had been revealed under X-ray. This crucial fact had somehow been overlooked in the scramble to get Tizard's box ready for its transatlantic crossing. The answer, as is so often the case, lay in a simple human error.

The Bell executives were completely satisfied, but now a strange conundrum revealed itself. Should the thirty magnetrons Bell had been contracted to produce be built with six cavities or eight? Should they follow the plan, or the equipment they had seen working and had in their hands? The answer, said Bowen, was straightforward—"to copy the one which they knew worked."[15] Simple it may have been, but it was also a tiny snapshot of the principles that would underlie the Allied attitude toward new technologies throughout the war—that getting it to work, and getting it into battle, trumped the perfect production process every time.

That potential hiccup overcome, Loomis's freight train powered onward. The following weekend the nucleus of his emerging microwave team reassembled at Tuxedo Park. Among them would be two newcomers—Frank Lewis, one of Loomis's own men who had been given the necessary security clearance to start work on microwave radar, and Ernest Lawrence, who had come from California to throw his considerable intellectual might behind the development of microwave radar. Initially Lawrence had been reluctant to get involved in war work. As a pure scientist, his research was above that sort of thing. But one person he did listen to was Loomis, who telephoned him in California to urge him to get involved. At this stage he could not, of course, be told details of the breakthrough that had been revealed to Loomis; but when Loomis's call was followed by a telegram from Vannevar Bush summoning him to the East Coast for an attachment to the Microwave Committee, he knew the time had come. Such was the persuasive power of Loomis and the project he was now overseeing, that Lawrence was only the first of a number of truly eminent scientists who began to recognize (or be convinced) that this work was of the utmost importance. Whatever their principal field of scientific endeavor, it now had to be put on hold while this new technology of war was pressed into service as soon as humanly possible. Lawrence, still basking in the intellectual glory of winning the Nobel Prize for physics the previous year, and others like him, had begun to see that this work was necessary not simply on its scientific merit, but as a means of preserving their ability to carry on their research in a free society. Microwave radar, the weapon "of the next war," was well on its way to becoming the weapon that would turn the tide of this one.

On the morning of October 10, a meeting was convened in the office of Karl Compton at MIT. The discussion, which went on deep into the afternoon, revolved around the idea that Loomis had conceived during that earlier weekend at Tuxedo Park when the British had revealed full details of the magnetron for the first time. It was only then that he had first fully appreciated the significance of the British use of civilian scientists in the development of radar. And only then that he decided to use the powers of the Microwave Committee to try to promote the establishment of a similar organization in the U.S., a special laboratory calling on all the skills and ingenuity of both academia and industry devoted to the development of microwave radar in all its forms. The next morning these ideas were put before the committee itself in a session held at the Carnegie Institute in Washington, and now the plan began to emerge in clear outline. It was to be a civilian research laboratory with a core of twelve scientists drawn from the top universities. That it would have to expand quickly was in little doubt. The possible applications of microwave radar in the theater of war were many and various. Getting them there would be a huge task—complex, time-consuming, and massively expensive. With Britain reeling under the Nazi onslaught in the air and at sea, there was a grave danger that technological salvation could come too late. So plans must be laid for quick and effective expansion once the lab got under way. From the nucleus of twelve, each scientist would be asked to put forward the names of four or five others, so that the recruitment process would not hold up a growing workload. One problem that remained, though, was where the lab might be physically sited. Whatever place was chosen would have to have good access to scientific facilities, plenty of room for growth and, crucially, as Eddie Bowen had pointed out, a nearby airfield with good hangaring—and aircraft too. There was also the issue of security to bear in mind.

Bowen lost to Ernest Lawrence at tennis that weekend at Tuxedo Park. By now Lawrence and Lewis had signed the oath of office that was a necessary requirement for working in the NDRC's radar committee, and had been brought up to date with developments. Loomis had asked everyone back to his upstate country residence, where the atmosphere seemed to create extra space for creative thinking. Bowen and Cockcroft, whose families back in the UK were now under nightly threat of death by high explosive, must have found it strange indeed to be enjoying themselves on a tennis court amidst the beauty of a Hudson Valley fall. On Sunday morning they got down to the job in hand. In a few short weeks Cockcroft would return to England, but Bowen's contribution to the defense of his country would be played out in this arena for months to come. For now, though, they were key to establishing the list of projects and the order of priority they would be given. They, unlike any of the other bright minds assembled at Tuxedo Park that weekend, had first-hand experience of modern war, and direct knowledge of what the British armed forces needed to defeat the Nazi threat. Loomis, according to Bowen, "paid us the compliment" of indicating that the priorities should be those of the British.[16] Working on those projects the new laboratory would have two goals—first to manufacture new weaponry to aid the British in their defense, and second to ensure that if war crossed the oceans to the United States itself, any aggressor would find it already armed with the very latest technology. There was no time to lose on either front.

Cockcroft would later write a succinct account of his part in the development of radar systems for the British Army. That weekend at Tuxedo Park stood out for him as a turning point:

> We had a memorable session with Loomis, Prof. Eddie Bowles, Frank Lewis, Ernest Lawrence, Bowen and myself. Loomis proposed the founding of a new microwave laboratory, the Radiation Laboratory. We discussed its programme and agreed that as night attack was England's direst peril the first objective should be a centimetric AI [airborne interception radar]. Bowen sketched the outline of the system and later took a great interest in its development. We suggested a centimetric G.L. [gun-laying] equipment—which was later started as the SCR-584. And from the U.S. side came proposals for a long wave pulse navigation system which later became LORAN. We were slightly embarrassed in these discussions since we had some knowledge of Gee which Tizard had decided not to disclose since it could easily be countered by jamming. I remember walking round the lake telling Loomis that we had a navigational system which could not at that time be disclosed. Within a few weeks from this meeting the Radiation Laboratory was in being—a laboratory which soon rivalled TRE in fertility and drive.[17]

The decision to create the Radiation Laboratory, later to become known as Rad Lab, was a major breakthrough. It was decided, from the start, that it was to be a civilian operation, with scientists and engineers drafted from the universities (on the British model), but also from industry—which would turn out to play a key part in the success of Rad Lab and its ability to get things done quickly. That it was, as Cockcroft said, up and running within "a few weeks" was largely down to the dynamism and contacts of Alfred Loomis. The Tizard mission had conveyed across the Atlantic not just that vital piece of hardware, but also, and just as important, Britain's sense of urgency forged by war.

Loomis was a man unaccustomed to waiting for the things he wanted. It was an attribute that would make him disliked in some quarters, but in time of war (even if not your own) it can be a highly desirable characteristic. Of the three priorities, airborne interception, gun-laying, and long-range navigation, the third was closest to Loomis's heart. It was the one area where the British were less than fully forthcoming. Loomis's interest in navigation stemmed from his love of yachting. He well understood the significance of being able to know exactly where you were on a vast and featureless sea. How much more vital that issue would become if you were abroad in a vast and empty sky—possibly at night, almost certainly with fighters and anti-aircraft guns waiting to bring you down. Add to that the responsibility of dropping bombs on to a designated target with, hopefully, pinpoint accuracy, a task which required you to know exactly where you were. Yet up till now this had proved remarkably difficult for aircraft.

This interest in the problem of navigation was one Loomis shared with Sir Henry Tizard, even though its source for Sir Henry was fundamentally different. Tizard had been of central importance back in the UK in promoting the use of radar as a defense against the bomber and of organizing the way it would be used. He did so because, in his capacity as chair of the Committee for the Scientific Study of Air Defence, he saw it as the only effective defense against the threat of bombing. But Tizard also headed the CSSAO, the committee studying air *offense*, and this double expertise gave him a unique insight into the possibilities of aerial bombing, but also its shortcomings. He knew, better than anyone outside a cockpit, how difficult it was for navigators to know precisely where they were, and therefore how important it was to develop some system of precision navigation. In GEE the British thought they had found such a thing.

GEE was in the very earliest stages of development in October 1940. In fact it had not yet acquired that name. GEE was an innovative use of pulsed radio transmission to allow a navigator to fix his place in the sky. Three radio stations, one a master (controlling the timings) and two slaves, spaced about a hundred miles from the master, would transmit synchronized signals to the target plane. The display equipment on board the aircraft showed the difference in reception time of the pulses, and thus the relative distance from the master and each slave. The aircraft carried a special GEE navigation chart with several hyperbolas plotted on it. Each hyperbola represented a line of constant time difference for the master and one slave station. All the navigator had to do was find the intersection of the two hyperbolas representing the two slave stations and then he would know his position with an unprecedented degree of accuracy. The beauty of GEE was that it could be used by any number of planes at one time, and that, vitally, the aircraft themselves had to carry no special transmitting equipment, which meant that a downed aircraft would reveal no secrets to an enemy. It had two severe limitations, though. First, it worked on a relatively short wavelength, so that it was limited to line-of-sight operation; and second, and as a consequence, the transmitting stations could not be too far apart. For aiding the navigation of planes over western Europe from transmitting stations in the UK it was fine. But beyond a couple of hundred miles, you were on your own. For offensive operations against Berlin or factories in eastern Germany, for instance, it was just about useless.

Tizard knew that some answer had to be found if Britain were to be able to bomb Germany and ensure its defeat in some (elusive) future. The difficulty lay in the fact that Britain was no place to develop a longer-range version of GEE for two reasons—one, the country was under attack, and two, it simply wasn't big enough. America, on the other hand, fitted the bill perfectly. Tizard's wish that the Americans might help in the development of some long-range system of navigation fell on willing ears in Alfred Loomis. In fact Loomis's keen interest in this particular need identified by the British stemmed from something more than just his knowledge and understanding of navigational issues. Loomis, a skilled electrical engineer, knew that the key to systems like this was synchronization; and the longer the range, the bigger that problem became.

The development of a long-range navigation system was right up there at the top of the Tizard commission's wish list. It was in the top three technologies set out by Cockcroft and Bowen on that October weekend in the luxurious surroundings of Tuxedo Park. It was perhaps not without some small sense of embarrassment that Bowen and Cockcroft brought this one up, because they knew that work had already started on the short-wave system, and that it would surely have been helpful to be able to talk about it freely. But they were not empowered to do that. Cockcroft was careful not to reveal anything about the new system, but Bowen, it seems, might have been a little more forthcoming.

Professor Ed Bowles was a key early appointment to the Microwave Committee. For some years he had been running microwave research at MIT, with a team of fifteen researchers. Bush and Compton had been keen that he get involved, but he and Loomis failed to hit it off right from the start. He saw Loomis as a manipulator and an autocrat, complaining later that he ran the committee as a personal fiefdom, reducing the other members to rubber-stamping his decisions or helping to fund-raise. It was the events of this weekend that began to move Bowles's uneasiness with Loomis into open dislike. Loomis and Eddie Bowen had taken to one another from the moment they met. Though they came from worlds apart, and differed considerably in age, they seemed to find much in common. This weekend, after the general outline of priorities had first come up for discussion, and Loomis had learned (in general terms only) of the existence of a prototype British system for radio navigation, he had gone out of his way to spend time alone with Bowen. Then it seems that, after everyone had disappeared to their respective rooms and come down dressed for dinner, Loomis announced that he had had a brainwave in the shower. According to Bowles he said he had suddenly seen exactly what would be required to develop a long-range navigation system. At this point Bowles claimed Bowen "winked" at him, as if to acknowledge that Loomis had been told a good deal more than the British were officially delegated to release. Bowles recalled of Loomis: "He took a great liking to Taffy Bowen, realizing here was a source to be cultivated. . . . It was clear that he was going to do his best to extract every vestige of information he could from Taffy on what the GEE system comprised. He was the kind of person who if one didn't know him, could seem to be operating very innocently. It was clear to me what he was up to."[18]

To Bowles this was a sophisticated and manipulative millionaire taking advantage of a naive young electrical engineer, though it is possible that Bowen knew exactly what Loomis was up to. Bowen's description of Loomis's "brainwave" is considerably more charitable, saying that Loomis had "fully appreciated" the importance of Tizard's perception of the need for a long-range navigation system and that "practically overnight" he had come up with doing something very similar to GEE "on the basis of the description I had given him of the British GEE."[19] Whatever the truth of the matter—who was manipulating whom, and just how much information Bowen passed on—Loomis's "brainwave" in the shower formed the basis of the LORAN system, which would be developed over the coming months as Rad Lab's Project 3. Meanwhile, Bowles eventually found himself removed from the Microwave Committee, while Bowen prospered at Rad Lab.

The creation of the basic specification for the other two projects that weekend is somewhat less contentious, but none the less significant. Cockcroft was the key man in setting down the parameters for Project 2. As an army man, with three years' experience in the Royal Artillery from 1915 to 1918, he knew exactly what the radar control of artillery could bring to the vital issue of range-finding, and he proceeded to outline on paper the specifications of a gun-laying radar set.

At the top of the list, what was to be Rad Lab's Project 1, was the design of an air interception set operating at the 10 cm wavelength. It would be christened AI 10. The first step, they decided, would be to design and build a prototype 10 centimeter system for detecting buildings at ground level. Bowen was central to the specification of the precise details of the design: "We listed desirable figures for things like pulse width, receiver sensitivity, cathode-ray tube displays and so on. On scratch pads and the back of envelopes, we sketched the block diagram of a typical system right there, with a modulator, a transmitter incorporating the magnetron, a receiver and indicator and appropriate power supplies."[20]

It must have seemed like science fiction. The potential unleashed by the cavity magnetron was allowing this handful of men to look into the future. On sketch pads and the backs of envelopes they were designing the next generation of weapons. In doing so they were shaping a new kind of war, envisaging a world where the bomber was suddenly exposed to a new threat of powerful night fighters that could see in the dark. Anti-aircraft guns would be able to track the bomber across a sky heavy with cloud. Coastal artillery would target ships twenty miles away to an accuracy of a few yards. New weapons would turn back the tide of advancing bombers and invasion fleets, and, once that task had been achieved, planes would find their way through darkness and overcast skies to launch their own assault on the aggressor with hitherto unheard-of precision. But as the pads and envelopes slowly filled with sketches, sums, and increasingly detailed specifications, they knew that this would all take time to achieve, and that it might all come too late. Detailed plans, prototype models, and production templates would all have to be designed and built before being put out to tender and production, and as yet even the Radiation Laboratory itself was still on the drawing board. Someone would have to instill the whole project with the urgency of war in a society still luxuriating in an uneasy peace. Alfred Lee Loomis was the man.

FOURTEEN

An American Mandate

THE FOLLOWING MORNING, OCTOBER 14, a small fleet of cars swept out of the gatehouse of Tuxedo Park and turned south toward New York City. Loomis had called a meeting in his penthouse apartment at 21 East 79 Street at 11 A.M. for as many of the Microwave Committee as could make it and representatives from the big electrical manufacturing firms. Westinghouse and GE had no one they could send at such short notice, but RCA and Sperry sent top men along. So, too, did Bell—but they already knew something of what was going on. Loomis launched straight in. He had gathered them all here because his committee had decided to begin construction of a ten-centimeter wavelength Airborne Interception system, AI 10. They were stunned. Even Bell Telephone's Mervin Kelly, whose labs were already commissioned to produce the magnetron itself, was taken aback at the speed of events. They would be yet more surprised by the time the meeting was over.

Bowen studied the slightly bemused faces of the industry executives as the magnitude of what they were being asked to do dawned on them. Loomis explained that he was looking for tenders for the manufacture of a wide range of components—magnets, pulse modulators, cathode ray tubes, receivers, and antennas. In thirty days. Loomis explained that despite the urgency of the project, it was still a competitive market. Sperry, Bell, and RCA all said they were able to take on the job immediately—others were clearly going to have to refer back to base. Those who said yes on the spot confirmed that they could deliver their tender well within the thirty-day time limit. "No," said Loomis, "you have misunderstood me. I want you to submit your tenders next week and to *deliver* thirty days after that."[1] The industry men were stunned. This was a speed beyond anything previously required in the military field. At this point Kelly was able to reveal that Bell's production of the magnetron was already in train, which was something of a spur to the competitive instincts of the other companies. GE decided they'd like to take on the job of producing the required magnet. Sperry put in for the parabolic reflectors and the scanning gear. RCA opted for the pulse modulator, power supply, and cathode ray tubes. (In fact Bendix would later win the bid to provide the power supply, and Westinghouse the antennae.) The bandwagon of microwave radar was up and running, and at quite some speed.

But where were these component parts going to be assembled into a single system that could be put through its paces? Where, in short, was the Microwave Committee's laboratory going to be? Who was going to work there, pushing the design and development forward? When the industry guests reeled out of Loomis's apartment that day, Bowen and Lawrence stayed behind. Now Loomis turned his persuasive powers on his old friend, just as the British had begun trawling universities for their brightest minds some few years earlier and Sir Henry Tizard had pulled in favors from his network of scientific connections. He wanted Lawrence to mastermind the recruitment process. Lawrence had clearly seen it coming, because he immediately volunteered two of his brightest students, Ed McMillan and Luis Alvarez. He had already opened tentative negotiations with his old friend Kenneth Bainbridge at Harvard. It was the first in a series of conversations with top scientific administrators at universities across the country through which the staff of the lab would be recruited.

But who would they be reporting to? Who would be the director of this cutting-edge laboratory—a laboratory still without a name or a home?

Home, for the moment, was Loomis's New York apartment, and it was there, on the morning of October 15, that Lawrence picked up the phone to talk to a brilliant young scientist then heading the department of physics at the University of Rochester. His name was Lee DuBridge, and he had been a protégé of Lawrence on his way to his current prestigious position. Despite the fact that Lawrence could tell him nothing over the phone of what the job entailed, beyond the fact that it was it was a vital position related to issues of national defense, DuBridge was on a train to New York City by nightfall. At thirty-nine years of age, he was about to become the director of the new Radiation Laboratory.

That same day London saw the heaviest bombing of the war to date. As another four hundred bodies were being pulled from the wreckage, there seemed no reason why this bombing should not go on for years, the skies over London virtually impossible to defend by night. Indeed, Churchill's belief that month was that London "would be gradually and soon reduced to a rubble-heap."[2]

In upstate New York, dislodged by the autumn breeze and the rush of DuBridge's passing train, gold and crimson leaves fluttered gently to the ground. The laboratory he was shortly to take over would have a major role to play in shifting the balance of power in the night skies over London, and in other arenas as yet unforeseen.

The rest of that week was spent in spreading the net of recruitment, while Bowen, Cockcroft, and other members of the mission shuttled between the labs of General Electric, Sperry, and Bell. Loomis and Lawrence, meanwhile, were tackling the tricky issue of where the new lab would be sited. Frank Jewett of Bell Labs (and the NDRC) suggested that his laboratories would be the ideal place. They had, after all, run a similar organization for the Navy during the previous war. Loomis had other ideas. Both he and Ed Bowles were former alumni of MIT, now run by another close associate, Karl Compton. Bush, too, had been dean of engineering at MIT a while back. They were confident Compton could be persuaded to house the lab, and MIT already had a strong track record in microwave research. East Boston airport was nearby, an ideal base for the experimental aircraft that would be needed. Also, in an environment like MIT the opening of a new lab would be perfectly normal, and might well attract less attention from outsiders. Jewett reluctantly agreed, and on October 17, they presented the idea to Compton, who had arrived for a meeting of the Microwave Committee. Faced with a fait accompli and some pretty powerful arguments, Compton got straight on the phone to his staff at MIT to see if they could somehow make available the necessary ten thousand square feet of lab space. It was the beginning of a new phase in the history of MIT.

On October 18, at the end of a frantic week, the most powerful scientific administrators in America reconvened in Washington, at the Carnegie Institution on P Street, from where Bush and Compton were running the NDRC. Mervin Kelly of Bell Labs was the only member of the Microwave Committee not present (he already had microwave work to attend to). Kelly's counterparts, the heads of research at GE and RCA, were there, as were Loomis, Bush, Compton, and representatives of both the Army and Navy. Loomis brought everyone up to speed with the decision to press ahead with centimetric radar. He told them what had been achieved so far, and that a new research laboratory was being created to push the project forward. He began to set out the tendering arrangements. The American electronics industry was about to demonstrate what it could achieve under the pressure of Loomis's all but impossible timeline. Bell stepped swiftly up to the plate—they could deliver five brand new magnetrons and crystal mixers inside the thirty-day limit. GE said they could produce the necessary magnet on the same timescale. Sperry could deliver the scanner as well as crystal mixers, but the klystron oscillators would take a little longer. Westinghouse said they would have a pulse modulator ready inside the month allowed; RCA was not so sure, and asked for ninety days instead. But they could promise a handful of cathode ray tubes inside thirty days. For Eddie Bowen this was the moment when the project really took flight, and it was after this meeting that "centimetric radar was on its way in the USA."[3] It was the perfect vindication of the decision to push the work forward through a civilian committee placing orders with commercial manufacturers.

By the end of October things were still gathering pace. On the twenty-fifth the NDRC allocated a sum of $455,000 to cover the lab's first year of operation. Space was already being cleared on the ground floor of MIT's main building and, after Bowen pointed out how useful the space on the roof at Bawdsey Manor had been in the early days, a large wooden structure of a thousand square feet was under construction on the roof. It would be the place where some of Rad Lab's most substantial breakthroughs were made in the coming months and years. Recruitment, too, was going well. When Columbia University's world-renowned physicist I. I. Rabi agreed to come over to the new laboratory, another future Nobel laureate was added to the pool of talent. Altogether ten members of Rad Lab's staff eventually would win science's top honor. At the end of the month Loomis sent Bowles off recruiting at a Boston conference on applied nuclear physics—six hundred upper-echelon scientists gathered in one building, rich pickings for a new and exciting project. Meanwhile Loomis and Compton hosted a lunch at the Algonquin Club, where a couple dozen of America's preeminent scientists sat down with Eddie Bowen, who had been smuggled in secretly through a back entrance. Although time was of the essence, anyone joining needed full FBI security clearance and would have to sign the Espionage Act—and they all had jobs to get out of. Nevertheless, a core staff of about twenty was lined up for the projected first day of operations on December 1.

But before that auspicious moment arrived there was the small matter of a presidential election on November 5. Over the previous couple of months the political positions of the candidates had begun to diverge. In Britain the nomination of Wendell Willkie as the Republican candidate at the end of June had been received with some relief. Willkie was a former Democrat who had been as much of an advocate of aid to Britain as Roosevelt. But since Labor Day Willkie had started to launch increasingly bitter attacks on Roosevelt, painting him as a warmonger who wanted to drag the U.S. into a conflict that it had nothing to do with. By the end of October the voters were looking at a much starker choice. Yet on the campaign trail Roosevelt altered his tone to placate isolationist sentiment. His promise at the end of October that "Your boys are not going to be sent into any foreign wars" was something of a body blow to the British, who saw FDR as a savior ready to come to the aid of Britain as soon as he had secured a fresh mandate. Churchill saw terrible dangers ahead should America decide to change its president: "No new-comer into power could possess or soon acquire the knowledge and experience of Franklin Roosevelt. No-one could equal his commanding gifts."[4]

Oliver Harvey, a British civil servant who had been private secretary to both Anthony Eden and Lord Halifax at the Foreign Office, described November 5, the upcoming U.S. election day, as "perhaps the most important date in the war. It will ring a bell throughout Europe. Our enemies will know that henceforward, even if they could succeed in wearing us down in a long war, there is a man across the Atlantic who would certainly bring America to our rescue, and would carry on the fight if we failed. A whole new continent ready to fight Hitler."[5]

Harvey's optimism was well placed. The mandate was massive—FDR won the popular vote by a margin of five million. It was a substantial victory, too, for the scientists putting together the new laboratory. The NDRC was a Roosevelt creation, and with it the Microwave Committee and the lab itself. Roosevelt's victory meant the Radiation Laboratory was still on track.

Meanwhile in the North Atlantic, a British convoy of thirty-seven ships, codenamed HX-84, was attacked by the powerful German pocket battleship *Admiral Scheer.* Escorted only by a single armed merchant cruiser, the *Jervis Bay*, the convoy had to scatter. Six ships including the *Jervis Bay* were sunk, forcing the British to suspend convoy sailings for almost a fortnight. Four hundred and forty sailors lost their lives that day, as the hopelessly outgunned *Jervis Bay* took on one of the most heavily armed warships afloat. As Americans went to the polls to turn their back on isolationism, the captain of the *Jervis Bay* was earning a posthumous Victoria Cross for valor displayed by staying at his post and fighting on after having an arm blown off by a shell from one of the *Admiral Scheer's* 11-inch guns.

In early November the register of recruits for the new laboratory was expanding rapidly. Rabi joined officially the day after FDR's victory at the polls. He soon brought with him two of his sharpest students from Columbia, Jerrold Zacharias and Norman Ramsey. From Harvard came J. Curry Street, Ivan Getting, and Edward Purcell. Luis Alvarez and Edwin McMillan caught the train from California in the second week of November, and November 11 saw the lab's first official meeting, a handful of men meeting in a makeshift office at MIT, where responsibilities were parceled out and teams selected. A week later Lawrence and Loomis drove up from Tuxedo Park with Bowen, who was pleased to see the wooden penthouse that had been built at his prompting on the roof of Building 6. He knew he'd be spending a lot of time there over the coming months. For Loomis this was a major shift in the focus of his activities. He'd already arranged for much of the equipment from the Tower House to be moved north, and today was the last act in transferring his personal center of operations to MIT.

Bowen's job that first day was to begin to pass on all his considerable experience by giving a lecture on military tactics and airborne interception. Loomis spelled out the lab's goals on a blackboard—among them the task of getting a microwave radar system up and running in the rooftop lab by January 6. This was to be followed by an actual airborne version installed in an airplane by February 1. Loomis clearly had the bit between his teeth. The next few weeks already promised to be exciting.

The team being brought together at MIT was among the brightest collection of scientists ever to be assembled in one place. They knew little, if anything, of radar at the beginning; but, as Luis Alvarez said, the real reason for them being there was that "they were the best people, and they were adaptable to anything."[6] They came from a background of international cooperation and awareness, and many of them were Jews or had close Jewish friends. They were acutely aware of what was going on in Europe, and grateful for the opportunity to try to do something about it.

And of course, there was Edward Murrow and his nightly broadcasts from a London reeling under the assault of the Blitz but, memorably, standing up and taking it. Murrow was trying in his broadcasts to give, he said, a sense of "the courage of the people; the flash and roar of the guns rolling down streets where much of the history of the English-speaking world has been made."[7] They were words that must have done much to encourage the bright young men now assembling in Cambridge, Massachusetts. What Murrow did not say, though, was that, as September rolled into October and Britain's anti-aircraft gunners raked the sky with their flak night after night, what those guns were doing was virtually useless. For every plane brought down, tens of thousands of shells were being launched into the night sky. More damage was caused by the falling debris. Microwave radar could not come a moment too soon.

Microwave radar, though, was still on the drawing-board, and its appearance in the theater of war was more than a year away. In the fall of 1940 no one had yet shown whether and how it could be made to work in action. But at the end of October a shipment from Britain arrived in Canada that would have an impact much sooner. The early days of the mission had been overshadowed by the decision taken in London *not* to send the ASV MkI developed and built by Bowen—a working airborne air-to-surface vessel radar set operating on a metric wavelength. Until proper trust was established it must have looked as though the British had something to hide. Now though, that omission had been put right with a vengeance, as the October shipment included the improved MkII version of the ASV set; an air-interception set, the AI MkIV; and an IFF unit. All of these were at the cutting edge of *practical* radar technology—stuff that worked—and Bowen now had the task of fitting them in airplanes to demonstrate to the American services what could already be achieved. And, of course, to whet their appetite for the new possibilities opened up by the cavity magnetron and microwave radar.

The month of November saw frantic work to install these systems, and by December all of them were airborne. At the Naval Research Laboratory two Canadians seconded from the NRC and the Royal Canadian Air Force got to work installing the ASV set into a U.S. Navy flying boat, after running it first on a test rig in the labs. When it took to the air for a test flight at the tail end of November the PBY Catalina was the first U.S. aircraft ever to be fitted with airborne radar. Bowen himself took his first flight on December 2.

Flying out over the Atlantic from Anacostia he was able to pick up an echo from a capital ship some sixty miles away. Bowen was elated when he saw what even this metric kit could achieve.

And it was badly needed. After the collapse of France and the end of Hitler's campaign in Norway, Europe had been subdued. Now the U-boats could be diverted to the North Atlantic to attack the convoys that were Britain's principal supply line. At the same time the Royal Navy's ability to protect convoys had been badly depleted by the need to take over the defense of the Mediterranean (and Empire supplies coming through the Suez Canal) from the French. After June 10, there was the Italian Navy to cope with as well. Suddenly the North Atlantic had become a very dangerous place. Between June and early November of 1940, over 270 merchant vessels and escorts were sunk. These six months came to be known by U-boat crews as *Die Glückliche Zeit*, the happy time. In September pack tactics were first used successfully by the German U-boats—to devastating effect. And, as the *Jervis Bay* and its convoy had discovered, German capital ships were now roaming the Atlantic too. But in the Catalina that day were the beginnings of a system that would ultimately defeat both the U-boat and the Kriegsmarine. So successful was the test that the Navy quickly ordered seven thousand copies of the ASV MkII from the Philco Corporation. Along with another ten thousand built in Canada, they would see active service over both the Atlantic and Pacific oceans and in virtually every theater of war.

Tied up as he was with the installation of the metric ASV set, with constant site visits and mission meetings, Bowen had little time to involve himself with another installation that was going on in parallel at the Army Air Force base at Wright Field, Ohio. This was the newly arrived AI MkIV, and the installation into a Douglas A20 Havoc light bomber was handled by Squadron Leader Hignett from the British Air Commission. Once again, the test flight proved highly successful, but the USAAF were not looking as far into the future as were the NDRC and the Microwave Committee. U.S. procurement decisions gave great weight to the defensive shield of the oceans in December 1940. Bombers seemed far less of a threat than they did in Europe, so the night fighter issue loomed nothing like as large as it did for the British. Twelve months later things would change dramatically; but for now only a token order was placed with Western Electric for the manufacture of AI MkIV radar, which would come into service as the SCR-540.

More immediately significant was the third system in the shipment sent from the UK, IFF, or Identification, Friend or Foe. Here again the mission brought across the Atlantic an invaluable lesson that had been learned in the heat of combat. For several months now day after day the skies over England had been filled with planes engaged in fierce, chaotic combat at over two, sometimes three hundred miles an hour. Crucial to the RAF's chances of containing the invaders, with their overwhelming advantage in numbers, was the ability of radar controllers on the ground not only to be able to pinpoint airplanes in the sky, but also to know whose aircraft they were. Nothing would have had a more detrimental effect on morale than fighters shooting each other down. IFF was essentially a system of secondary radar developed by the British where an RAF plane picked up by radar would transmit a signal identifying it as a friendly aircraft, enabling the operations staff to sort out their own fighters from enemy raiders on the plots in front of them. In November an example of this system arrived in the U.S. and was fitted to a U.S. Navy fighter at the Naval Research Laboratory, working to a range of ninety miles on its first outing. The benefits were immediately clear, and with little fuss the system was adopted for use by both the U.S. Navy and Army Air Corps as their standard means of identification. The adoption of a shared system of identification with the British was a powerful though tacit statement of joint interest, and a vital step toward the successful combined operations that were to characterize the Allied airborne war effort in the years to come.

Back in Britain November was a grim month. The day before Americans went to the polls Londoners had woken up to a strange feeling. They had just enjoyed their first uninterrupted night's sleep for almost two months, and if Roosevelt won the election, they believed the U.S. would surely come into the war and the tide would begin to turn. But the initial elation at Roosevelt's victory was followed by disappointment, even disillusionment, when the U.S. failed to appear over the horizon (despite having initiated the draft on October 15). The lull of a night without bombs turned out to be short lived as the wail of the air-raid sirens returned—followed by the dreadful drone of bombers overhead. For ten nights the bombers came, dropping their deadly load on London in a seemingly unbroken stream. On the fourteenth they attacked the provincial city of Coventry and its auto plants instead; then it was three more nights of raids on London. But suddenly the German strategy changed, and London became just one target of many. Now all the industrial cities in Britain were legitimate targets for the Luftwaffe, and the whole of Britain was to experience at first hand what those in the capital had been enduring for the previous ten weeks. During the two months that followed there would be twenty-two raids of more than a hundred bombers strong against major industrial towns all across Britain.

If Britons were disappointed that America stayed neutral after Roosevelt's crushing victory over the Republicans, there was nevertheless some encouragement to be taken from unfolding events across the Atlantic. With a new four-year mandate the tenor of FDR's speeches began to change. On December 17, he used his regular press conference to set out a new theory of promoting American self-interest through support for Britain at war:

> I go back to the idea that one thing that is necessary for American national defense is additional productive facilities; and the more we increase those facilities—factories, shipbuilding ways, munitions plants, etc., and so on—the stronger American national defense is. Now orders from Great Britain are therefore a tremendous asset to American national defense, because they create, automatically, additional facilities. I am talking selfishly, from the American point of view—nothing else.[8]

Selling to the British, Roosevelt said, was a way to build up American defense—at someone else's cost. The trouble was, the British were finding it increasingly difficult to pay for the matériel they so desperately needed—they were running out of dollars.

When, in November 1939, the United States had replaced the outright prohibition on arms sales to belligerents with the principle of "cash and carry," the Royal Navy was in complete command of the Atlantic and was successfully maintaining a blockade of the German merchant fleet. This meant that the British could buy armaments from the U.S., while Germany was effectively prevented from doing so. Britain's difficulty at the close of 1940 lay in just how successful that policy had been. Payment for these arms had to be made in dollars. Britain had entered its war with reserves of $4.5 billion in dollars and gold, together with some U.S. investments that could be turned into dollars. Over the first fifteen months of war sales of goods (often luxury goods) to America had brought in an additional $2 billion, but the rate of arms purchase had soared, and more or less doubled when French orders were taken over after the collapse of the Republic. By November 1940, Britain had already spent $4.5 billion buying arms from America, and the bottom of the barrel was now distinctly visible.

Nine days before Roosevelt's December 17, 1940, press conference Churchill had dispatched a letter to Roosevelt setting out the parlous position Britain now found itself in. Lord Halifax and the Foreign Office, along with Ambassador Lothian, had wanted the letter to concentrate on the country's financial predicament. Churchill, though, was determined to stress the centrality of what he was shortly to characterize as the Battle of the Atlantic to Britain's survival. He insisted that he must focus on the practicalities of the dangers facing shipping on the North Atlantic, where U-boats were now able to operate deep into the ocean from their new bases in occupied France. Lothian and Halifax had to give way to Churchill and his experience of personal correspondence with Roosevelt; but they were pleased with the final version. The four-thousand-word letter that was finally dispatched to Roosevelt on December 8 summarized the events of the previous twelve months and set out the grim prospects for 1941. Britain now needed three things from the U.S.: arms (and aircraft in particular); financial aid, without which she could simply not afford them; and American shipping of these vital supplies and policing of the Atlantic sea-lanes. He urged Roosevelt to use "every ton of merchant shipping" he had and to support them with U.S. Navy escorts. Britain's three interlocking needs were starkly spelled out. It was time for Roosevelt to act. "Mr President," concluded Churchill, "this is a statement of the minimum action necessary to the achievement of our common purpose." Churchill would describe this letter as the most important he had ever written.[9]

Roosevelt received this letter on board the USS *Tuscaloosa*, an American warship on which he was taking a well-earned rest cruising the Caribbean. Immune from the opinions of all but his closest advisers, he pondered the letter for two days, rereading it several times as he basked in the sunshine in his deck chair. That Britain was running out of money was not news to him. Indeed, the issue of what would happen when it did had been raised in the cabinet immediately after the election. About this time Oscar S. Cox, a smart lawyer in the Treasury Department, had, after prompting from Secretary Henry Morgenthau, unearthed an 1892 statute declaring that the Secretary of War could, "when in his discretion it will be for the public good," lease Army property (if not required for public use) for a period of not longer than five years.[10] It was the germ of an idea that Roosevelt ruminated upon as he sat on the deck of the *Tuscaloosa* turning Churchill's letter over in his hands.

Until the collapse of France, Britain had been desperately trying to balance its dollar budget. Faced with the prospect of imminent invasion, Churchill's government had decided that arms had to be ordered immediately, and the financial consequences worried about later. Now, after some six months, unfulfilled British orders (yet to be paid for) amounted to some $2.6 billion. And a further order had been placed that just about doubled that amount. The position was unsustainable. Roosevelt had to find a solution.

The *Tuscaloosa* docked at Charleston on December 14. Three days later, just before his scheduled press conference, the president lunched with Treasury Secretary Morgenthau. He laid out before him a plan that had been conceived on board the *Tuscaloosa*: "I have been thinking very hard on this trip about what we should do for England, and it seems to me that the thing to do is to get away from a dollar sign. I don't want to put things in terms of dollars or loans, and I think the thing to do is to say that we will manufacture what we need, and the first thing we will do is to increase our productivity, and then we will say to England, 'We will give you the guns and the ships that you need, provided that when the war is over you will return to us in kind the guns and the ships that we have loaned to you, or you will return to us the ships repaired and pay us, always in kind, to make up for the depreciation.'" Morgenthau said he approved of the idea, and added: "If I followed my own heart, I would say, 'Let's give it to them'; but I think it would be much better for you to be in a position that you are insisting before Congress and the people of the United States to get ship for ship when the war is over, and have Congress say that you are too tough, and say, 'Well, let's give it to them,' than to have the reverse true and have Congress say you are too easy."

From these events—Lord Lothian's promptings, Churchill's long letter, a newly elected president taking a break in the Caribbean, and a lunch with a trusted lieutenant—the extraordinary policy of Lend-Lease was born. By the time of that press conference on December 17, the full import of Churchill's letter had filtered through into administration policy, and it was now spelled out to the American public in Roosevelt's uniquely accessible style. Picking up from his assertion that building for Britain was also building for America's defense, Roosevelt moved on to the problem of the financing of Britain's future orders. A way had to be found (a way acceptable to the public) of facilitating the purchase of arms without appearing to be taking a step closer to joining Britain at war. He couched it in the sort of homey terms that had come to characterize his fireside chats on the radio: "Suppose my neighbor's home catches fire, and I have got a length of garden hose four or five hundred feet away; but, my heaven, if he can take my garden hose and connect it up with his hydrant, I may help him to put out his fire. . . . I don't say to him before that operation, 'Neighbor, my garden hose cost me $15; you have got to pay me $15 for it.' . . . I don't want $15—I want my garden hose back after the fire is over."[12]

The plan was that the United States would take over British orders, turn them into American orders, and then either lease or mortgage the matériel to Britain: "on the general theory that it may still prove true that the best defense of Great Britain is the best defense of the United States, and therefore that they would be more useful to the defense of the United States if they were used in Great Britain than if they were kept in storage here."[13]

It was the outline of a policy that would underpin Britain's lone stand against the Nazis for many months to come. If America itself was not coming into the war, then at least maybe its factories were.

Lord Lothian, Britain's prime mover in shaping this policy and author of much that went into Churchill's letter, did not hear of Roosevelt's "garden hose" speech. He fell ill suddenly on December 12 and died, after refusing to consult doctors in accordance with his religious beliefs as a Christian Scientist. He left his mark in both Lend-Lease and the benefits soon to begin flowing from the Tizard mission that he had done much to shepherd through.

Christmas came and went with little sense of cheer on the streets of London and Britain's other big cities. The bombs that had been dropping on London since early September were a mixture of high explosives and incendiaries. The high explosive aerial bombs had been supplemented by a tactic of dropping much bigger naval mines by parachute. Not only were they larger, but their destructive power was further enhanced because they would explode at ground level instead of driving themselves into the ground on impact. Sometimes, however, they would fail to explode on contact, leaving the city dotted with unexploded mines lying in gardens or stuck on railings, or lodged in the attic of a house—each one with the destructive power to sink a ship. And each one discovered meant evacuating up to a thousand people while specialist teams tried to defuse the mine, all the while listening for the distinctive buzzing that meant you had fifteen seconds to run for your life. This was going on night after night.

So far the incendiaries had proved easier to cope with than at first thought. The principal weapon was the thermite incendiary, about eighteen inches long and weighing under a kilogram, or just about two pounds, so that a bomber could carry thousands of them, often in canisters that would burst to scatter incendiaries like a modern cluster-bomb. They were at their most dangerous when they fell into empty offices or warehouses or through the roofs of houses with their windows blacked out, when they could start fires not noticed until they had taken a firm hold. If they fell on to a road or a pavement, though, they could be left to burn out, and it wasn't difficult to extinguish them with a blanket or drop them into buckets of water. But in December all that changed, when the Germans started to fit incendiaries with a small charge of high explosive, so that anyone trying to deal with them would be in effect picking up a live hand grenade. The war was becoming dirtier yet, and there was always that fear of what new weapons the Germans might bring to bear; there were rumors of a "radium bomb," early fears about nuclear fission research.[14]

The radium bomb never came, but on December 29, the Germans struck lucky. The initial raid that night was small by comparison with many that London had endured before. A second wave of bombers never came, because the weather closed in to make flying impossible. But the damage done by incendiaries dropped in the first raid was to outstrip anything that had gone before: 136 bombers came that night, dropping 127 tons of high explosive bombs and 613 canisters of incendiaries. (This was not unusual in the winter of 1940—five times this number had been dropped on the night of December 8.)

Fire-watchers were few on the ground, it being Christmas week and a Sunday. The City of London was fertile ground for the incendiaries that night. When they came, the firemen soon found themselves confronted by an abnormally large number of burning buildings, with more and more as the evening wore on as small, unseen fires grew into major blazes. A 50 mph westerly wind fanned the flames. Then a high explosive bomb fractured the city's principal water main, a twenty-four-inch pipe running from the River Thames to the Grand Junction Canal. Pressure from the pumps driving the firemen's hoses suddenly dropped to virtually nothing, and powerful jets were reduced to a trickle. Such an eventuality had been foreseen, though, and emergency pumps had been installed along the banks of the Thames to draw water from London's arterial river to help in its defense. The Thames, though, is still a tidal river as it passes through London. That night, the tide was out and with the river at its lowest ebb, the feed pipes of the pumps were simply unable to reach the water. The Germans had timed the raid specifically to coincide with this low tide. With water supplies virtually at zero the firefighters could do little but watch their city go up in flames.

By the time the German raiders left at about nine o'clock in the evening there were over 1,400 separate fires in the central City district alone. Quickly they formed into two huge conflagrations, the larger of which consumed everything in an area half a mile square just to the north and east of St. Paul's Cathedral. When it had burned itself out the blaze left the biggest area of war devastation in all of Britain. The day after the raid the BBC broadcast one man's description of the scene around the cathedral:

> For miles around the city was a bright orange-red—the balloons in the barrage [overhead] stood out as clearly as on a sunny day. St. Paul's Cathedral was the pivot of the main fire. All around it the flames were leaping up into the sky. And there the Cathedral stood, magnificently firm, untouched in the very centre of all this destruction. As I walked along the streets it was almost impossible to believe that these fires could be subdued. I was walking between solid walls of fire. Groups of shops and office buildings came down with a roaring crash. Panes of glass were cracking everywhere from the heat, and every street was criss-crossed with innumerable lengths of hose.[15]

The survival of London's St. Paul's cathedral, Sir Christopher Wren's masterpiece, built to replace a predecessor destroyed in the Great Fire of London in 1666, was something totemic for the British—if St. Paul's could survive such a conflagration raging all around it, then maybe somehow Britain could survive too. The majestic dome, rising three hundred and fifty feet above the smoldering ashes, stood out as a beacon of hope. On that same day, December 29, came news of another such beacon—this one from across the Atlantic. In his last fireside chat of that momentous year, Franklin Roosevelt pledged America's support to Britain by helping supply the matériel with which it could fight the Nazis. What was clear from the very start, though, was that Roosevelt had no intention of taking America to war: "My friends, this is not a fireside chat on war. It is a talk on national security; because the nub of the whole purpose of your President is to keep you now, and your children later, and your grandchildren much later, out of a last-ditch war for the preservation of American independence and all the things that American independence means to you and to me and to ours." He was keen to stress, too, that this was not what was being asked of America, though in fact these sentiments would scarcely have been echoed on the other side of the Atlantic. The people of Europe, he said, "do not ask us to do their fighting. They ask us for the implements of war, the planes, the tanks, the guns, the freighters which will enable them to fight for their liberty and for our security. Emphatically we must get these weapons to them, get them to them in sufficient volume and quickly enough, so that we and our children will be saved the agony and suffering of war which others have had to endure."[16]

This was not, though, simply a matter of getting involved in somebody else's fight. What was happening across the Atlantic had direct relevance for Americans gathered around their radios listening to his speech.

> The Nazi masters of Germany have made it clear that they intend not only to dominate all life and thought in their own country, but also to enslave the whole of Europe, and then to use the resources of Europe to dominate the rest of the world. It was only three weeks ago their leader stated this: "There are two worlds that stand opposed to each other." And then in defiant reply to his opponents, he said this: "Others are correct when they say: With this world we cannot ever reconcile ourselves. . . . I can beat any other power in the world." So said the leader of the Nazis. In other words, the Axis not merely admits but proclaims that there can be no ultimate peace between their philosophy of government and our philosophy of government.[17]

This, then, was a major public statement of the United States's community of interest with Britain as the sole surviving democracy in Europe. A public commitment of what Roosevelt had been working privately toward for some time, the diversion of a considerable portion of America's productive capacity to the manufacture of goods for one of the principal belligerents in a foreign war. "We must be the great arsenal of democracy. For us this is an emergency as serious as war itself. We must apply ourselves to our task with the same resolution, the same sense of urgency, the same spirit of patriotism and sacrifice as we would show were we at war."[18]

This speech was very far from the declaration of war that many in Europe had hoped would come from his lips before that terrible, blood-soaked year of 1940 was out; but it was a very substantial step along that road, and no matter how many times Roosevelt emphasized that Americans would not be entering the battlefields, or climbing into their cockpits, the import of this speech was as clear to Berlin as it was to London. Roosevelt's speech had at least given Londoners something to hang on to, a hint that things might change in 1941.

FIFTEEN

Radiation Laboratory

In December 1940, the Loomis family assembled as usual at their 17,000-acre Hilton Head Island hunting and fishing retreat. Ellen and Alfred's three sons were all away on duty with the military, and Eddie Bowen, only a little older than the eldest, found himself treated almost like a fourth son. Karl Compton and his wife had traveled down, too, and Bowen was looked after royally. Bowen was a long way from his family, including his wife Vesta and their young sons back in South Wales. Now that the Luftwaffe was spreading its attentions beyond London to Britain's other big industrial cities, South Wales, a coal-mining area with huge steel works, was under major threat. It was an anxious Christmas for both Eddie and his wife, separated by war and thousands of miles of ocean. He was spending Christmas in a completely different world—Loomis's world of luxury and privilege a stark contrast to Bowen's recent postings at bare-bones military research outposts around the UK. It must have been a very strange few days for him, but by early January they were back at work.

That winter was cold in Massachusetts. Miserable, too, with plenty of rain and snow to dampen the spirits in the drafty and unheated rooftop lab at MIT. The young scientists hard at work building their prototype pulsed microwave radar set were driven by the deadline of January 6, the date by which they had to have the system working. Eddie Bowen was spending more and more time at the laboratory. His frantic round of meetings and site visits had begun to die down, and his expertise was in great demand in Cambridge. Bowen was of a similar age to many of the Rad Lab team, but he already had far more experience. He had been working on airborne radar for nearly five years, he had seen radar in wartime action at first hand, and he brought with him the precious cavity magnetron. Small wonder, then, that he was somewhat revered by the men of Rad Lab.

With such a tight deadline the work was relentless. Fresh faces seemed to be arriving on almost a daily basis—each one already eminent in some field or another. New recruits would take a room in the Commodore Hotel just off Harvard Square while they found a place to stay, and, where necessary, made arrangements for their families to travel to New England. What little recreational time there was, was generally spent in the bar of the Commodore. Bowen remembered those evenings with great fondness, noting that "there was always a good turn-out and the hotel did a roaring trade." It was an ironic place to celebrate this extraordinarily open cooperation between Britain and the United States, because the hotel bar was decorated with a mural depicting episodes from the American struggle for independence from Britain—the Boston Tea Party, Paul Revere's Ride, and the Minute Men. Bowen recalled, "Proceedings usually began with an expression of solidarity, a friendly toast to the British—'The hell with the Limeys.' It was all in good humour and there were many friendly barbs about certain failings of British tactics in the battles which finally lost them a big slice of the American continent."[1] These Friday night sessions became known as Project 4, and despite, or perhaps because of, them good progress was being made on the other three.

On the first Saturday of 1941, January 4, the roof-mounted microwave set was fired up. Less than eight weeks had passed since the first men walked through the door of the new laboratory, and already they had a functioning prototype. This was the day the system picked up its first echoes, from the tower of the Christian Science church across the Charles River basin in Boston. The trouble was, the equipment was unwieldy—a long way from something you could mount in an airplane. It had been built using a separate transmitter and receiver, while the minimum requirement for something compact enough to fit into a fighter was that a single antenna should do both jobs. This meant using a duplexer of the sort that had been under development at the Naval Research Laboratory at Anacostia. The duplexer worked by switching off the sensitive receiver during the powerful transmission pulses that would otherwise burn out the receiving crystal. It was a simple enough concept, but one they were finding hard to put into practice when the transmitted pulse was something like a million times stronger than the faint echo bouncing back from a distant object. "For the first, and possibly the only, time," wrote Bowen, "a mood of pessimism crept into the group and some doubts were even expressed about whether the system would ever be capable of receiving echoes from an aircraft."[2] DuBridge set several teams to work on the problem, using the British principle, adopted by Loomis, of putting a functioning solution ahead of purity of design. It was cracked by a team headed by Jim Lawson, an amateur radio enthusiast, who came up with a system using a klystron amplifier as a buffer between the antenna and the receiver crystal. It was a start, but they were still having trouble in picking up signals by scanning. A signal could be obtained by pointing the transmitter directly at a known target, but this rather missed the point. Back in Washington, Vannevar Bush was piling on the pressure. He needed results to fend off the damaging attacks being mounted by some industrialists, who were deeply skeptical of the ability of this band of seeming dilettantes to come up with practical designs that could be put into production.

Bowen, with typically British understatement, described it as a "rather frustrating period," but it was mercifully short; DuBridge's teams were able to iron out the remaining problems with scanning, and on February 7 the system, now with a fully functioning duplexer, picked up its first echoes from an aircraft flying over East Boston Airport at a range of some four to five miles. Not quite the "actual airborne version installed in an airplane by 1 February" that Loomis wanted, but nonetheless substantial progress. Meeting in Washington that day, the Microwave Committee was in desperate need of something to show for its efforts, and it was with considerable relief that DuBridge took the phone call to tell them that the system had at last successfully picked up an airborne signal. It was enough to silence the critics on the committee, which gave its blessing to the laboratory's pressing on with Project 1.

Early in January the Lend-Lease Bill was introduced to Congress in the face of considerable opposition. The former ambassador to the United Kingdom Joseph Kennedy and Charles Lindbergh were among the bill's most prominent opponents. On that same day, January 10, 1941, pacts were signed between Germany and Russia on frontiers in Eastern Europe and on trade. Germany was now guaranteed a steady supply of food and raw materials in exchange for industrial equipment.

By February merchant convoys transporting goods to Britain across the Atlantic were increasingly vulnerable as the German battle-cruisers *Scharnhorst* and *Gneisenau* were now loose in the Atlantic, along with the heavy cruiser *Admiral Hipper*. In response to the growing threat there was a major reorganization of the U.S. Navy into three fleets, the Atlantic, the Pacific, and the Asiatic. Admiral Ernest King was given command of the new Atlantic Fleet, which would be a significant strengthening of U.S. forces in the Atlantic. British naval forces, meanwhile, were heavily engaged in the Mediterranean, and, a bright shaft of light in the surrounding gloom, XIII Corps, the Western Desert Force under General Archibald Wavell, had made major inroads against the Italians in North Africa. Since the start of the campaign two months previously, Wavell, with no more than two divisions, had destroyed ten Italian divisions and taken 130,000 prisoners for the loss of 555 dead and 1,400 wounded.

The pace at which the conflict was unfolding seemed to be relentless, and as the cold winter winds whistled around their rooftop laboratory Rad Lab's young scientists knew that there was no time to lose in the race to perfect their new weapon systems if they were to have an effect on the war's outcome. Now they had a system that could detect an airplane in flight; the next step was to reengineer it so that it could be made to fit in a fighter. The job was handed to Luis "Al" Alvarez and Ed McMillan, who started by building a wooden mockup of a Douglas B-18's bombardier's compartment, the type that was coming to them for their first airborne tests. It took them three weeks to get the installation right, including putting in a special Plexiglas screen that would allow microwaves to pass through. In early March the B-18 was delivered from Wright Field up to Boston airport by an Army Air Corps crew, and by March 10, Rad Lab's first microwave AI set was ready to take to the air. Extensive ground testing had shown it to be pretty reliable, and McMillan, who was in charge of the installation, was confident of success as the B-18 took off. That confidence was not misplaced. For the first time they were able to detect ground targets using airborne microwave radar. A little fewer than 90 days had passed since the doors of Rad Lab opened. What they didn't know was that on the other side of the Atlantic a similar flight was taking place that day with an even more satisfying result.

Tizard had long ago identified the threat that would eventually come from night bombing if and when the British radar-assisted fighters managed to fend off the Luftwaffe's daytime raids. That time had now come, but thanks to Tizard's foresight, much work had been done to try to develop battle-ready airborne interception systems on both meter-wave and centimeter systems. The AI MkIV which had been shipped out to the U.S. in November of the previous year was one of the first combat-ready fruits of this work. In fact as Bowen was returning from Wembley clutching the precious magnetron bound for America, the final adjustments were already being made to a Bristol Beaufighter fitted with AI MkIV at the RAF's Fighter Interception Unit at Tangmere in southern England. The AI MkIV operated on a metric wavelength. It was old technology, but it worked.

The Beaufighter was a new aircraft with far better speed, firepower, and maneuverability than its predecessor, the Blenheim. The MkIV finally resembled the production-ready AI system that would be needed if it were to make any difference in the battle for air supremacy. Eddie Bowen had had a sneak preview of the Beaufighter in the summer of 1940. He'd been flown down to the factory at Bristol by Jumbo Ashfield, one of Fighter Interception's top pilots, to see the airplane being built. He was mightily impressed:

> The Beaufighter is an aircraft which radiates an enormous aura of power—both engine power and fire-power. The internal layout seemed ideal for a radar-equipped aircraft and Jumbo and I looked in awe at the powerful Hercules engines and four 20 mm cannon nestling in the belly. With four to six more .303s in the wings, this was indeed a formidable machine. We headed for home with much lighter hearts. A Beaufighter equipped with the latest AI radar clearly had the potential to make an enormous difference to the coming night battles.[3]

The MkIV-equipped Beaufighter R2055, with Jumbo Ashfield at the controls, made its first flight on August 12, 1940—just as Tizard was preparing to leave for America. Plans were set back a little when the plane was damaged during a German attack on Tangmere, but it was soon repaired and back in action. Its performance was clearly a big step forward, and on the night of September 4–5, it flew its first operational sortie. Just over a week later, however, the difficulties of developing these systems in Britain in the heat of war were thrown into high relief when R2055 was lost in action off the French coast. Britain's most advanced air-interception aircraft had lasted precisely one month.

There were others to follow, of course, and by the first week of November 1940, the AI MkIV was back in action. Despite that early setback the system promised much, particularly now that it could be used in conjunction with the new ground-controlled interception (GCI) radar coming through at the same time. Ground controllers could now locate the incoming raider and guide the Beaufighter into the right position to be able to pick up the bomber with its own onboard radar. Even in the dead of night. Now came a gift from the gods—a seven-week lull in the bombing onslaught. Persistent bad weather over Britain from mid-January to early March largely kept Germany's bombers on the ground. It was a welcome respite for the beleaguered population of London, and it could not have come at a better time for the RAF. They went into it with old-fashioned, lightly armed, and underpowered Blenheims as their principal defense against night bombing, and came out of it with radar-equipped Beaufighters and GCI.

But, revolutionary as they were, AI MkIV and early GCI were still first-generation radar, operating on a long wavelength, with all the concomitant problems of lack of precision and ground return swamping the signal. Quite simply, the radar was only good over a distance equivalent to the height at which the aircraft was traveling—a distinct drawback against low-flying raiders. Back at the TRE laboratories in Swanage work was well advanced on the next generation of equipment—using the cavity magnetron to build the precision radar of the future. In fact the work had been proceeding pretty much in parallel with developments at Rad Lab, so much so that in early March airborne microwave radar took to the skies on two continents on the same day. March 10 marked not only the first flight of Alvarez and McMillan's B-18, but also the first airborne outing of TRE's centimeter kit, called the MkVII. The British had had a bit of head start, and on that day they were still in front, because the system developed by Alan Hodgkin at Swanage, installed now in an old Blenheim, was already being tested, not on a ground target, but against an airborne one, which they were able to pick up on their indicator screen at a range of some two to three miles. It was the beginning of a constructive rivalry between the two labs that would give the Allies a crucial advantage over their enemy in years to come.

But these were only proving flights, and operational microwave AI, with all its benefits of accuracy and capability of use at any height, was still some way in the future. As if to emphasize the urgent need for it, on March 8, two days before airborne microwave radar took to the air on either side of the Atlantic, just when Londoners had begun to dream the worst might be over, the weather cleared and heavy bombers appeared again in the skies over London. That night 120 bombers were in action over the city, and they would continue almost without interruption until the middle of May.

Before March was out the Rad Lab team lifted off in their B-18 to test the equipment for the first time against an airborne target. It was March 27, and the plane was full—with McMillan, Alvarez, Ernie Pollard, and Bowen all on board. Their target plane was a single-engine machine supplied by the National Guard, and its instructions were to fly out into clear air over Cape Cod Bay at about 10,000 feet. The B-18 made several runs on the target as it criss-crossed the bay, with excellent results. They found they could pick up the target at ranges well in excess of two miles, which ground return would have made impossible using meter-wave gear. It was a clear demonstration of the superiority of centimetric radar even in its prototype form. They intended to turn for home at this point, but Bowen, remembering his success at Bawdsey in tracking a Royal Navy Task Force on maneuvers in the English Channel, decided to test the new system's ability to pick up ships. To plot the precise position of an airplane in the sky the system had to take three readings—fixing the distance, or range, was a simple matter of the time it took for the signal (traveling out at the speed of light) to return; detecting where the target was in the sky was achieved by scanning in azimuth (side to side), which would give the target's bearing, and elevation (up and down), to find its height. Switching off the elevation scan allowed them to sweep the surface of the sea looking for ships. They quickly picked up clear echoes from a ship of about 10,000 tons traversing the bay, at a range of about ten miles. It was an impressive result for a prototype system, and spurred Bowen to try for something extra.

Just about half an hour's flying time to the west of Cape Cod Bay, at New London in Connecticut, lay one of the U.S. Navy's principal submarine bases. Bowen reasoned that, having demonstrated their ability to detect both airplanes in flight and large ships at sea, it was worth taking the chance to see how the kit would perform against a submarine. There was no guarantee, of course, that any vessels would actually be at sea but, as luck would have it, they found a large U.S. Navy submarine cruising and fully surfaced. It was a sitting duck of a target, and several broadside runs showed they could pick up a surfaced sub at a range of four to five miles. Flushed with success, but low on fuel, they banked the B-18 and headed for home. It was the first time airborne radar had successfully picked up a submarine, and this success would do much to affect the future direction of research.

Back in the fall of 1940 airborne interception had been given top priority because it was at the top of the wish list of the visiting British, then locked in the Battle of Britain. But now that the first results of Rad Lab and the Tizard mission were coming through, the situation had changed markedly. Britain had succeeded in repelling the attempt to knock out its air force, but was still laboring under aerial assault from the bombers. Now a new danger was becoming ever more evident as Churchill emphasized to Roosevelt in his letter of December 8—the increased threat to the Atlantic shipping lanes posed by the new U-boat bases on the coast of occupied France. German U-boats were now operating freely in the Atlantic without having to run the gauntlet of the Royal Navy's blockade of the North Sea. And they were moving closer than ever to the eastern seaboard of the United States.

In the second half of 1940 U-boats had sunk the vast majority of the 315 ships lost in the Atlantic. The convoy system offered some protection, but with the appearance of many of the big ships of the German surface fleet in the Atlantic it was increasingly necessary to provide battleship or cruiser escorts, and these were already spread thin. Moreover, Royal Navy warships had to steam from British ports to escort the convoys, so the U-boats had begun to move further and further west, picking up the convoys in mid-ocean when their escorts had to turn back.

The threat to American commerce in the Atlantic was becoming ever more obvious, just as America itself seemed to become ever more entangled in Europe's war. American factories were put to work building ships, arms, and aircraft. In the heavy seas of the Atlantic the nature of America's future war was growing clearer. The threat of the U-boat was a major concern to the U.S. Navy, and the Rad Lab team's success in detecting a submarine came as welcome news. Adapting the AI system developed for Project 1 into a purpose-built ASV set was a comparatively simple matter, and the Navy quickly placed an order for a trial system.

March had been a big month for the Radiation Laboratory on the administrative front too. Such had been the success of the recruiting campaign that after four months of existence the lab now had nearly two hundred people on the payroll, more than half of them front-line scientists and engineers. Lee DuBridge had brought in an associate director, F. Wheeler Loomis (no relation of Alfred Loomis), to run the administrative side of lab business. I. I. Rabi had been placed in charge of core research and he had created subgroups to concentrate on things like increasing the power output of the magnetron and starting development of the next generation microwave apparatus, working at 3 centimeters instead of 10 centimeters. So far the lab was meeting all its targets. Work on the core projects was bringing out more potential applications of the new technology, and the range of development just kept on growing.

All of this, though, was running well beyond what had originally been envisaged. Alfred Loomis and the Microwave Committee submitted their first report on the Radiation Laboratory to Bush and the NDRC at the beginning of March 1941. It was a detailed account of almost unmitigated success, but it was also a plea for more funding: $300,000 was needed immediately, and an estimated $1 million to cover the salary bill for a further year. The tricky part was, though, that Rad Lab had been established without any guaranteed source of funding, because the NDRC had to get Congressional approval to release the funds. Congress had dragged its feet, to the point that Rad Lab was on the verge of running out of money. Something had to be done quickly to prevent work grinding to a halt. It was time for Loomis and Compton to call in favors, and soon they were able to pass on the good news that the salary shortfall was going to be covered by an anonymous guarantor.[4] Congress, though, spurred by the imminent threat of actual war, voted the funds through and the philanthropic offer of cash was not needed. According to some accounts the anonymous donor was America's richest man, John D.

Rockefeller Jr.; but it seems far more likely that Loomis himself took the risk. Whatever the answer, the Rad Lab express was kept firmly on the rails, and work began on finding a new structure that would guarantee the funding in future.

Project 1 was already well advanced, and beginning to spawn further projects for use against ships and submarines, and by the end of January, Project 2, the search for radar-guided anti-aircraft gunnery, was starting to take shape. So significant did the British regard this, that they initiated parallel research efforts in the UK and Canada. Each of those systems would later come to play a significant part in the war, but the system produced at Rad Lab would revolutionize anti-aircraft defense.

The problem of hitting a bomber flying fast and high was well known, and in the fall of 1940 it was reckoned that the success rate was somewhere around 30,000 shells fired for every plane brought down. Conventional radar, of course, was a help, and anti-aircraft guns were already being equipped with meter-wave radar, which could pick up the course of an airplane through overcast conditions or at night. But it was still not precise enough. Microwave radar, however, held out the prospect of much greater precision; and it was this that Rad Lab and the other teams in the UK and Canada were asked to work on.

Project 2 was perhaps the most extraordinary example of radical thought in the early days of Rad Lab. Existing meter-wave sets for gun-laying were a tremendous improvement on what went before, but all they did was to indicate the airplane's position—gun-laying, or aiming, was still performed manually. The 10 centimeter microwave radar offered the possibility of plotting the position of the target plane with much more accuracy, but of itself could do little to improve aiming. In 1941 all three teams came up with working microwave systems, and it was the Canadians who got there first. Their system, the GL3C, was, however, fairly crude. The British GL3B was better, but still suffered from the problems of manual aiming. On both British and Canadian systems it was the job of a human being to move and aim both the radar antenna and the gun itself. The Rad Lab team decided to go for something far more ambitious. They reckoned that centimetric radar would only really come into its own if they could develop a gun-laying set that would lock on to and automatically track the aircraft, even through evasive maneuvers, and then transmit that information to an automatic aiming device. It was a bold and daunting task; but such was the verve of the Rad Lab team that they immediately set to work on what would be a quantum shift in anti-aircraft gunnery—if they could pull it off.

Alfred Loomis's broad range of scientific interests served the Microwave Committee and Rad Lab very well, and never more so than now. Loomis was a member of both the Royal and American Astronomical Societies, and his thorough grounding in astronomy was the basis for his early suggestion that the precision of microwave radar might be significantly enhanced by the use of a conical scan.[5] The idea was to move the transmitter in a circle so that the "cone" it was radiating was itself rotating around a central axis, such that the axis would always be kept inside the cone but was effectively moving within it. Theoretically this would enable the target to be held on that central axis, because the revolving cone would detect any movement off the axis and, crucially, the direction of that movement.

A target located on the axis will generate a constant return no matter where the cone is pointing, but if it is off to one side it will generate a strong return when the cone points in that direction and a weak one when it is pointing away. It should, therefore, be comparatively easy to make the radar adjust automatically, and even to pass that information to servos powering the gun's movement that will automatically track the position of the radar to aim the gun itself. That was the plan. And they were in the right place. MIT had a Servomechanisms Lab that was at the leading edge of developments in servo control, the new field of automatic devices that used error-sensing feedback to correct their performance.

This was a potential breakthrough of considerable proportions, and DuBridge assigned the project to one of his brightest men—Louis Ridenour, a brilliant physicist from the University of Pennsylvania. A small team was assembled around him, and in

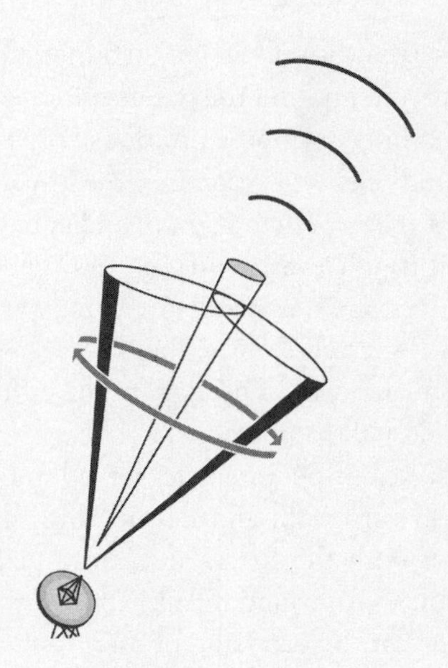

Conical scanning solved the problem of acquiring targets with the accuracy necessary for gun direction. The beam of a microwave radar is adjusted to be slightly off center. By rotating the radar, the beams will form a cone that overlaps at the center. A target that is picked up by this radar will have a weaker return on the outer edges and a progressively stronger one toward the center. Once a target enters the narrow beam of overlapping signals, it can then be tracked by keeping the target within that beam. In the SCR-584, the radar signal is fed to a director computer which then provides this information to the gun.

January 1941, they began their work. There was a further considerable problem to overcome in trying to keep the system from becoming confused with extraneous signals bouncing back from other objects caught in the rotating cone—trees, or buildings, or, most likely, other aircraft. This, though, was a problem that a team of physicists was ideally placed to solve. They were able to borrow another concept from the world of astronomy by using a highly accurate quartz crystal clock, triggered as each pulse was transmitted, to time a pair of traps to open at precisely the right microsecond to allow through only a signal coming from the precise distance of the locked-on target, thereby screening out all extraneous signals.

By the end of April, Ridenour's team had a prototype radar system up and running. To test the automatic aiming system, they mounted the radar onto a General Electric automatic gun turret destined to be installed in a Boeing B-29 bomber; they removed the guns and replaced them with a 16 millimeter camera. For a friendly "target" they conscripted a friend of one of the team to fly his light plane around the area for $10 an hour; by May 31, the system was able to track the aircraft accurately. It was a remarkable achievement in such a short space of time, and work quickly moved on to making the system suitable for use in the field, mounting the whole set-up on a single trailer with the antenna on top. Known as XT-1, the system was first tested at Fort Monroe in February 1942. Its performance was outstanding. It was able to detect bombers at a range of up to forty miles, and could automatically track them from about eighteen miles, at which point it was accurate down to about twenty-five yards in range, and 0.06 arcseconds in angle.

But there were further refinements yet to come. In parallel with the development of the XT-1, Bell Labs began working on an analog computer that would be able to direct the guns using electrical instead of mechanical input. It would be known as the SCR-584 and, in conjunction with Bell Labs's analog computer, the M-9 Director, was demonstrated in its complete form on April 1, 1942. Orders for 1,200 systems were placed the next day by the Army, and the SCR-584, a radically advanced design produced in quantity in America's factories, would become one of the most significant weapon systems of World War II.

When, on the new AI set's first proving flight against an airborne target on March 27, 1941, Eddie Bowen had asked the pilot of the B-18 to carry on along the coast to New London instead of heading home, one man on board was particularly interested by their ability to detect ships at sea. Ernest C. Pollard was a Briton who had been working in U.S. labs for quite some time, and with loyalties straddling the Atlantic he was an ideal recruit to Rad Lab. Pollard spent his early childhood in China. At the age of ten he moved back to the UK when his father died. After completing his schooling in England he went on to study physics in Rutherford's Cavendish Laboratory at Cambridge University. In 1932, he took a Ph.D. under James Chadwick, who had recently discovered a previously unknown particle in the atomic nucleus. The discovery came to be known as the neutron because of its lack of an electrical charge, and it would win him the Nobel Prize in 1935. As Pollard set out on his Ph.D. it was already clear he was being supervised by a physicist at the very forefront of his discipline. (Chadwick's discovery was vital to the fission of uranium 235, and the later development of the atomic bomb.) But

Pollard's future was to be in America. In 1933 he travelled across the Atlantic to take up a position at the physics department of Yale University, and there he designed the university's first cyclotron in 1939. It was a resumé that made him a perfect candidate for Rad Lab.

Pollard joined the new project at the beginning of 1941 and was given charge of a team developing naval applications of the cavity magnetron. Soon after the lab's first experimental 10 centimeter kit had been demonstrated, the Navy had requested that Rad Lab set to work on a centimetric search radar for shipboard use. Britain had tested the prototype of its first centimetric radar set, Type 271, aboard the corvette *Orchis* in March 1941. By May 1941, a high-definition, 50 kilowatt, 10 centimeter radar set was ready for sea trials on board the U.S. destroyer USS *Semmes*. It was ultimately given the designation SG, and, like Type 271, it would soon be demonstrating its worth.

Shipborne radar, however, presents some unique problems. First of all, the ships are moving. Land-based anti-aircraft gunnery suffers from the difficulty of a tiny target moving at high speed many thousands of feet above. But at least it sits on a stable platform. Shipboard radar sits on a platform about as unstable as they come. Ships roll and pitch and yaw, they twist and turn through the sea, sometimes deliberately to make themselves a more difficult target for attacking aircraft or U-boats, but also because they are being tossed about by random patterns of wind, wave, and swell. Their speed, too, constantly changes. So the job of radar on board ship is immensely complex, as it has to take all this motion into account to have any chance of giving an accurate reading. In the spring of 1941, these were intractable problems, and it would be about a year before SG was ready to be installed in production form. But the trials on the USS *Semmes* that May held out great hope.

The defense of ships in the Battle of the Atlantic, though, was by no means simply a matter of the ships' ability to defend themselves. Hunting down the U-boats from the air offered the chance of destroying the menace before it got anywhere near a convoy. By early 1941, Bowen's early-design airborne radar fitted to the airplanes of RAF Coastal Command was already beginning to have an effect. Once again, it also served to demonstrate what could be achieved with the greater accuracy and precision of microwaves. On April 29,

Rad Lab received a distinguished visitor from across the Atlantic, on what Eddie Bowen would recall as a "red letter day." Air Chief Marshal Sir Hugh Dowding was the man who, as head of the RAF's Fighter Command, had won the Battle of Britain. Now heading a purchasing mission to the U.S. and tasked with equipping the RAF for the battles to come, he understood better then anyone the urgent need of a precision airborne interception radar.

April 16 and 19 saw two more huge bombing raids on London. Londoners referred to them simply as "the Wednesday" and "the Saturday." A broadcast announcement of the attack on German radio claimed that as many as 100,000 fire bombs had been used on the Wednesday attack alone. Air Chief Marshal Sir Charles Portal, Chief of the Air Staff, issued a public message congratulating Fighter Command's night fighters on their splendid performance against the enemy night bombers. The squadron had brought down nine German raiders in the space of three nights, four of them falling to a single pilot who would later be awarded the Distinguished Flying Cross for his work. First-generation airborne interception radar was proving its worth. Now Dowding had come to America to see what the Rad Lab team could offer by way of improving the system.

Just about two years had passed since Bowen had last had the chance to demonstrate the latest AI system to Dowding, two years in which the world had changed dramatically. When they took off for Dowding's first demonstration flight at Suffolk in the summer of 1939 war was a looming shadow on the horizon; now it was a bloody reality. Dowding's presence in America was tangible evidence of the new culture of communication spawned by both the Tizard mission and the strong bond created between Roosevelt and Churchill. Bowen observed that Dowding was "surprised to find that things were so advanced" and keen to see for himself what Rad Lab's microwave system could achieve.[6] Lee DuBridge invited him to join Bowen, Ed McMillan, and himself on a proving flight in the B-18. It did not disappoint.

After lifting off and climbing to 10,000 feet above the Charles River basin and Boston Harbor they were soon able to pick up the target aircraft at a distance of some five to six miles. The value of microwave radar was instantly apparent. At this height the signal of the conventional AI set now in action in the skies over Britain would have been drowned out by ground return. The new system could see three times as far—picking out tiny aircraft, in darkness or cloud, at the unheard of range of more than five miles. It must have seemed like manna from heaven to the man responsible for the night fighters, though it's doubtful the scientists could pick it up from the poker face of the notoriously phlegmatic "Stuffy" Dowding.

On his return to Washington, Dowding immediately wrote to the U.S. Army Air Corps to report on the outstanding performance of the new equipment, and urged them to do all they could to make sure that the system was put into production as soon as possible. Once back in the UK he made sure to place his own order through the Ministry of Aircraft Production:

> To: British Air Commission
> From: Ministry of Aircraft Production
> MAP 6557 9/6 Despatched P.S.3 9.6.41
>
> Have been much impressed with the results obtained with NDRC Laboratory 10cm AI. Should much appreciate immediate despatch of ten such sets to this country by air, with 200 to follow subject to such modifications as we consider desirable after obtaining experience with the first ten sets.

Back home in the UK, the TRE had reached an advanced stage with the development of its own 10 centimeter AI set, but this order placed in Washington for the Rad Lab design, to be manufactured by Western Electric (Bell Labs's manufacturing arm), was the first practical embodiment of Tizard's belief that America's vast productive power could be called into action in the war against fascism.

In the U.S., however, military enthusiasm for AI radar remained low. It was the new naval implementations and the development of the SCR-584 anti-aircraft gunnery system that were attracting all the interest. Trials of the SG shipboard system on the USS *Semmes* had continued to produce impressive results, and early in June Ernie

Pollard, on board the destroyer for the trials, had been able to demonstrate the system's capabilities as a navigational aid. A thick fog had descended quickly on the ship as it was in mid-exercise with a handful of submarines. Normal procedures required the captain to drop anchor and stay put until the fog cleared. Pollard, though, was able to demonstrate to the captain that the SG could pick out the harbor buoys rendered invisible by the fog. It was little short of miraculous for the captain to be able to see, for the first time, the position of the buoys on the plan position indicator (PPI) screen, and instead of being immobilized by the fog, he was able to bring the *Semmes* and its accompanying submarines safely back to harbor despite hopeless visibility. The Navy's production order followed swiftly—the first American order for 10 centimeter radar.

But in the UK microwave airborne interception remained high on the military wish list, and the order that followed Dowding's visit was clearly only scratching the surface of what the Rad Lab boys felt they could offer. So plans were laid to take the Rad Lab prototype across the Atlantic so that it could demonstrate its worth in practical trials. The new Douglas A-20, an airplane that was being shipped over to Britain to be fitted with the old-technology AI MkIV, was proving to be well suited to the night fighter role. Known in Britain as the Havoc, it seemed a perfect platform for the new system; but unfortunately the RAF delivery schedule meant that the Rad Lab men simply couldn't get their hands on one quickly enough.

It was the Royal Canadian Air Force who came to the rescue, with the offer of a civilian Boeing 247D that they had been using for their own experimental work. The airplane was flown down to Logan Field, where the installation was handled by a bright new recruit named Dale Corson. Once the new 10 cm AI had been fitted to the Boeing, it was flown down to Newark, New Jersey, then crated up for shipping to the UK as deck cargo. On June 28, 1941, Eddie Bowen returned to the UK along with Dale Corson. They were flown over on a B-24 of the North Atlantic Air Bridge, ready to receive their specially equipped Boeing 247D. To Bowen it was tangible proof of the mission's success: "This was just ten months after the Tizard Mission had left Liverpool in August 1940. The 10cm magnetron had been transported to the U.S. and was already being manufactured in large quantities. The Radiation Laboratory had been formed and was in full operation and a wide variety of centimetric radars were being manufactured. In less than one year, one of these sets was delivered to Britain, already installed in its own aircraft."[7]

Unfortunately, Bowen's positive vision of the Boeing visit was not shared by everyone in Britain. The plane was flown to Tangmere to be put through its paces for the specialist Fighter Interception Unit. By June 1941, the worst of the German night-bombing attacks was over—and meter-wave AI fitted in Beaufighters had played a major part in that. So, while the pilots of the Fighter Interception Unit were all too aware of the importance of AI radar, they had grown understandably attached to the system they already knew—a system that had performed so well that the Germans seemed to have gone home with their tail between their legs. While the unit produced a report which was full of praise for the new equipment, no one at Tangmere seemed to feel any urgency to replace the existing system.

Bowen's old colleagues at TRE were even more lukewarm. They had been busy developing their own 10 centimeter system, and, while they might have been keen on America's manufacturing capability, they clearly had scant respect for the Americans' ability to come up with a decent design. It was a classic case of "not invented here." Nevertheless, comparative tests were carried out, and it turned out that an interesting anomaly had arisen. When Bowen, Cockcroft, and the others had embarked on the *Duchess of Richmond* all those months earlier they had taken the cavity magnetron, enabling the Americans to develop the powerful transmitter they lacked. The Americans, though, had been way ahead of the British in receiver technology. But by the summer of 1941, Bowen was surprised to find that that position had been reversed—here he was demonstrating to the British a transmission system he considered superior, while the British, clearly having worked hard on their weaknesses, now had the better receiver technology. Each group, he concluded, "had been striving to improve the weakest link in their chain and each had made considerable progress. I like to think that both sides profited from the exchange."[8]

That they did so was in large part due to the third group to whom the American system was now demonstrated, officers at the very top of Fighter Command—Dowding's men. They were mightily impressed by what they saw, and were acutely aware of the parlous position in which the RAF still found itself, in respect of both the ability to withstand renewed German night-time offensives, and their growing dependence on America for supplies of hardware—from bombers to bullets to new technology. The equipment in the Boeing was, after all, just work in progress, and if the Americans could come up with the best airborne interception system, Fighter Command would be eager to get their hands on it.

After three months back in Britain demonstrating the American microwave radar and liaising with old colleagues at TRE, in October Bowen returned to a United States still at peace. Still at peace, but seemingly becoming more entangled in Europe's war as the months rolled on. The U-boats were now operating within five hundred miles of the Canadian coast. In April, American warships and aircraft had begun regular sweeps along the trade routes up the East Coast and out into the Western Atlantic. Their orders were to report and shadow any Axis ships or submarines they located. They were under strict instructions not to engage in any offensive action, but it was nevertheless a policy guaranteed to outrage the Axis powers, and quite likely to provoke the sort of incident that Roosevelt knew would be the only way the majority of American public opinion could be swung behind the U.S. entry into the war. Hitler's attention, however, was focused elsewhere, as his Axis allies would soon discover.

On May 10, London was again battered by a huge air raid—the biggest yet in terms of casualties. More than five hundred bombers attacked the city during the night, inflicting terrible damage on some of the British capital's most iconic and important buildings. The British Museum and the Royal Palace at St. James's were both badly damaged. The seat of Britain's political power, the chamber of the House of Commons, was virtually destroyed. Miraculously, the magnificent medieval Westminster Hall alongside it emerged almost unscathed. Winston Churchill was photographed picking his way defiantly through the wreckage of the Parliament building in an attempt to bolster the spirits of the London public, though 1,364 people were killed that night and more than fifteen hundred seriously injured.

It was a devastating raid; but mysteriously the next night the skies over London were quiet. And the night after that. Six days after the destruction of the House of Commons a hundred bombers attacked Birmingham, Britain's second-largest city; but again the skies fell silent. This was the last major air raid on a British city for many months to come. By the middle of May the Luftwaffe had seemingly withdrawn from the fray—and they were not alone. In the Atlantic there was a sharp decline in the number of U-boat attacks. They, too, seemed to be required elsewhere. They would be back, of course, but in the meantime they had left behind them a vital gift to the Allied war effort.

The day before London's Houses of Parliament were reduced to rubble, U-boat U110 was harrying Atlantic convoy OB-318 outward-bound from Liverpool. The German submarine had already sunk two ships when it was forced to surface by a depth-charge attack. The stricken U-boat was boarded by seamen from HMS *Bulldog*, one of the convoy's escorting destroyers. In heavy seas the boarding party soon found themselves stranded on the enemy vessel when their small boat, tied up alongside the U-boat, began to break up in the waves. They quickly collected together all the documents and equipment that they could find before the *Bulldog* managed to pluck them off. The crew did not know it for years to come, but they had captured one of the most significant pieces of equipment in the German armory, the Enigma coding machine (together with its vital documentation) used by the Kriegsmarine to encode the messages to U-boats that had made their hunting so effective. For the moment the U-boat menace had eased; but when they returned they would be facing radar-equipped ships and planes, and an enemy that seemed to know their every move. Three weeks into June it suddenly became apparent where the Luftwaffe and the U-boats had gone.

SIXTEEN

A Two-Ocean War

On June 22, 1941, more than four million Axis troops poured across the borders into Russia along a front stretching 1,800 miles from Lithuania south to the Black Sea. Hitler had ripped up his nonaggression pact with Russia to launch Operation Barbarossa, the invasion of his neighbor to the east, making good the promises and threats of *Mein Kampf.* Now Britain had an unexpected ally, and an extended commitment in the Atlantic. Immediately after Russia came under attack Churchill pledged to send all possible aid to its defense against the Nazis. This meant an extension of the Atlantic convoys with the opening of a new route round the top of Norway to the north Russian ports of Archangel and Murmansk. Loaded in Britain or America the convoys would assemble in Iceland for the perilous voyage through the Arctic Ocean. Iceland was now becoming a pivotal point in the supply lines to the Allies, and in July the United States took another tentative step closer to the European theater of conflict in the North Atlantic. The British had occupied Iceland in May of the preceding year following the German invasion of Denmark, at that stage responsible for Iceland's foreign affairs. It was a preemptive strike to prevent this strategically positioned island from falling into German hands. On July 7, 1941, responsibility for the defense of Iceland was transferred to the still-neutral United States, by mutual agreement with the Icelandic government. American troops arrived to replace Britain's garrison of 25,000, now urgently needed elsewhere.

Major moves were afoot back in the U.S. on the technical front too. The difficulties over the financial shortfall had prompted a review of the status of the NDRC. Thus far the organization had been dependent on special funding from the Oval Office. Vannevar Bush now felt that this was holding them back, and that the time had come for funding to be put on a more solid basis. In May he set out his plans to the Bureau of the Budget. He wanted to extend the brief of his organization beyond the research stage and into the creation of actual prototypes. It would put the development of fresh military hardware squarely in the hands of scientists instead of the military. The idea fell on fertile ground at the White House, and in June Roosevelt signed an executive order to upgrade the NDRC to a new entity, the Office of Scientific Research and Development (OSRD). It would be run by Bush and could apply directly to Congress for funding.[1] Weapons development was now freed from the immediate predilections of the military to become more future-oriented.

One of Bush's priorities was to build on the early reciprocal exchange established as a direct result of Tizard's mission. In the winter of 1940, Harvard University President James Conant had traveled to the UK to establish an NDRC liaison office. The British had reciprocated by opening a bureau in Washington run by the mathematical physicist Charles Darwin, grandson of the great man whose name he shared. Bush was keen to see that these high-level exchanges were mirrored by actual scientists crossing the Atlantic. Among the first to travel to Britain was a young physicist named Norman Ramsey, who arrived in mid-June to begin the reverse flow of technology stimulated by the Tizard mission. Ramsey was the leader of Rad Lab's Advance Development Group, whose task was to experiment at the very forefront of the available technology. To this end they decided to push the magnetron to its limits and go beyond the 10 centimeter range in search of even better resolution. On the roof of the laboratory's Building 6 they had put together an experimental 3 centimeter system, and it was this development that he was now carrying across the Atlantic. He was to spend several weeks traveling in the UK to discuss his 3 centimeter work as well as watching night fighters in action from a Chain Home station on the east coast.

Britain was almost like home to Ramsey—just a few years before he had been completing a second undergraduate degree in physics at the Cavendish Laboratory in Cambridge. Many of the top men at TRE, and at Oliphant's laboratory in Birmingham, were familiar faces. This was to stand Ramsey in good stead as he helped organize the comparative tests of the AI system brought over by Bowen for demonstration in the Boeing 247D. The far superior performance of the British receivers had come as a big shock to Ramsey and the rest of the American contingent now assembled in Britain. For a while they tweaked their own receiver to try to coax more performance from it, but soon gave in and coupled a British receiver with their own (superior) 10 centimeter transmitter. The results delivered by this hybrid British/American set were impressive—aircraft could suddenly be detected at three times the range. It was a dramatic demonstration of what could be achieved with the sharing of information.

Early American attempts to build receivers had been based on using crystals (a bit like the ones in an old "cat's-whisker" type of crystal radio receiver), but they had soon been abandoned in favor of vacuum tubes. The British, however, had stuck with the use of crystals throughout, and had apparently made some significant breakthroughs. Oliphant's lab had come up with a design with much better resistance to the shock and vibration that had made early models impractical for use in the field. The British had also made a major advance in the duplexing problem with their receivers. The perennial difficulty of protecting the receiver from the surge of power from the transmitter had been solved with the invention of the Sutton Tube, which temporarily cut out the receiver during the transmission phase by creating a short circuit in a low pressure gas that was ionized by the transmitted pulse. As soon as the transmission stopped, the gas recovered and the circuit was reestablished. It was a solution as elegant as it was effective. When Ramsey returned to the U.S. he would take these advances with him. With him went TRE's second attachment to Rad Lab, Denis Robinson, the man who had kick-started the British work on crystal receivers.

Ramsey returned to America as a convert to the crystal receiver; and he set about trying to spread the word to all the companies with major microwave radar development programs. It has been called the first "silicon public relations tour."[2] Within a matter of months Bell Labs, Westinghouse, and Sylvania Electric Products were all working on major crystal research and production programs. Purdue University and the University of Pennsylvania weighed in too. At Rad Lab a project was initiated to improve crystal reception by experimenting with the introduction of impurities that might enhance conductivity. A program was now under way to develop the use of silicon crystals that would soon begin to shape the future of electronics through the rest of the war and beyond.

Robinson arrived in the U.S. with a message. Project 1 was no longer top of the list of priorities. With the withdrawal of the Luftwaffe for Operation Barbarossa, and with metric radar performing well against the remaining bombers, the Battle of the Atlantic and the fight against the U-boat now became the top priority. Robinson told the Microwave Committee that microwave air to surface vessel radar should go right to the top of the pile. At the beginning of 1941 the menace of the U-boats had started to increase at an alarming rate. At the outbreak of the European war large numbers of new submarines had been commissioned by the Kriegsmarine, and now they were just coming into service. They came in two standard sizes—500 tons and 740 tons. Their range was considerable, 11,000 miles and 15,000 miles respectively. Soon they were to be joined by "U-cruisers" whose range at slow speed could reach 25,000 miles. These were formidable weapons once they were roaming free in the Atlantic. The opening months of 1941 brought exceptionally bad weather in the Atlantic. Heavy seas caused many unsuitable vessels to lose their place in convoys to become the "stragglers" that were such easy pickings for the U-boats now operating in the deep ocean. During the first two months of the year fifty-seven ships had been sunk by U-boats. Then, in March, the Germans introduced their fearsome "wolf pack" tactics. Now, once a U-boat picked up a convoy, instead of attacking it would call up other U-boats, then simply shadow the convoy, hiding in its wake until the full pack arrived and they could attack in numbers. It did not help that there was a gap in the mid-Atlantic where no air-cover could be provided due to the limited range of aircraft. The losses of shipping began to rise alarmingly.

In response to this growing threat Churchill issued a Battle of the Atlantic Directive aimed at building up convoys' defenses. Merchant ships were to be fitted with catapult-launched planes and anti-aircraft guns as quickly as possible, and instructions were given that more RAF Coastal Command squadrons were to be formed and fitted with the new airborne radar systems as soon as possible.

Protecting the convoys required massive coordination of effort, with destroyer escorts operating out of both the UK and Iceland, and the RAF flying in close support of convoys or sweeping the Atlantic in the hunt for patrolling U-boats. In the month of April alone Coastal Command aircraft flew a million miles supporting convoys and 200,000 miles on anti-submarine patrol. The weak point in the armory, however, was that the aircraft could achieve very little even when they found themselves in the vicinity of a U-boat. In daylight a surfaced submarine would almost certainly see the aircraft before being seen. It could then crash dive and alter course below the surface, becoming instantly undetectable from the air. At night the U-boats were essentially invisible—that is, until the arrival of Coastal Command aircraft fitted with ASV in early 1941. Its immediate impact was succinctly summarized in a joint report on the scientific war published in 1945 by the Office of Scientific Research and Development, the War Department, and the Navy.

> The great effectiveness of radar in the war against the U-boats arises precisely because of the nature of the submarine. A submarine relies on submersion for concealment, but it cannot live indefinitely under water. In the days before radar, it could get along very well, even in hostile waters, by proceeding submerged during the day and surfacing to charge its batteries and take on fresh air during the night; its chances of being detected in the darkness were small. There is no night in radar's world, and when the ability of radar to see a submarine for some ten miles is combined with the ability of a high-speed airplane to travel some hundreds of miles in an hour, a

> systematic sweeping of the areas of the sea in which submarines are known or expected to be becomes practicable. This is what the RAF Coastal Command undertook to do.[3]

On May 9, the old policy of using aircraft to accompany convoys was switched to direct attack, using the new technology to hunt U-boats in areas where they were most likely to congregate. By the late spring the meter-wave ASV MkII set developed by Bowen's group at St. Athan had been fitted to more than a hundred aircraft. If this system could be combined with the use of very long-range aircraft, it might even be possible to bridge the mid-Atlantic gap in defenses. The ultimate goal, of course, would be to fit such an airplane with a microwave ASV set instead of the MkII. These were the things Denis Robinson was asked to pursue when he got to Rad Lab.

At the end of May the MkII system demonstrated its worth against surface ships when a radar-equipped Catalina located the battleship *Bismarck* heading for repairs in the shipyard at Brest. It was limping to port after sustaining serious damage during the action in the Denmark Straits in which it had just delivered a hammer blow to the Royal Navy, sinking the totemic battleship HMS *Hood*. Guided by the Catalina, Swordfish torpedo planes from the carrier HMS *Ark Royal* homed in on the stricken *Bismarck* in heavy overcast, one torpedo delivering a crippling blow that left Germany's most prestigious naval weapon a sitting duck for the British warships who closed in to finish her off. It was a high-profile prize for a fledgling (and still secret) system. (The official version publicized at the time was that the Catalina sighted the *Bismarck* through gaps in the cloud.) But this success masked the limitations of meter-wave radar against the much smaller target of an often partially submerged submarine. A microwave ASV system would change everything.

Denis Robinson had more than one reason to be pleased he had arrived in the U.S. It was a major personal opportunity, but it was also a chance to reunite his young family. Several months earlier, after they had narrowly escaped being killed in a German air raid on Swanage, Robinson had shipped his wife and two young sons out to New England to stay with his brother, a Harvard academic. For a few precious months they would all be together again, safely out of reach of the German bombers. And Robinson would be able to add the invaluable wartime experience of TRE to the productive power of Rad Lab.

No sooner had he arrived than he spelled out to Lee DuBridge the reasoning behind the British request that priority be given to the submarine problem. DuBridge decided Robinson should speak directly to Alfred Loomis, the lab's chief policy-maker as head of the Microwave Committee. Robinson flew down to Washington and was immediately invited to Loomis's suite at the Wardman Park, the same rooms where the cavity magnetron had first been revealed. Now, less than a year later, Robinson had the difficult task of persuading the Americans that the electrical engineering powerhouse they had created over the past nine months would have to make a drastic change in course. He need not have worried.

Arriving at Loomis's rooms for the meeting, Robinson was amazed at the extent of American informality. The issue was discussed over a large glass of whiskey. Far from being put out at the suggestion, Loomis was completely unflustered by the British request. He told Robinson that Rad Lab would go along with whatever they wanted, and "tell DuBridge I said so."[4] It was almost as if Loomis was expecting it.

In fact when Robinson got back to MIT to discuss the matter with DuBridge himself, it soon became apparent that the two men had talked about the issue in advance of the Wardman Park bourbon. The response they had decided upon said much about the "can do" approach that underpinned the entire undertaking. They saw no reason to stop work on airborne interception in favor of the new priority work on ASV. "We've got the power; we've got the people. We'll do everything," said DuBridge. AI work was to continue, but Robinson was to be given a free hand to set up a stand-alone ASV project sharing an office with one of Rad Lab's most accomplished scientists, I. I. Rabi. Any doubts Robinson might have had about their commitment to the new goal were swiftly set aside as the equipment he needed soon began to pour in to the laboratory. "I'd never seen so much technology available," said Robinson. "There wasn't any shortage."[5]

Before leaving England, Robinson had been charged with finding out which of the new long-range bombers now beginning to arrive in

Britain under the terms of the Lend-Lease Act would be most suited to be fitted with anti-submarine radar. Such an airplane was the weapon that could fill the mid-Atlantic gap. He had chosen the B-24 Liberator, and he soon found himself with two of these new aircraft at his disposal at East Boston Airport, along with a commitment from Rad Lab to get them fitted with microwave ASV as soon as possible. "When I told DuBridge what I'd been sent over to do, he said, 'Fine, we'll arrange that—we'll make ten of these things for you here in Radiation Lab.' They were called DMS 1000. . . . I was given complete authority to have it designed the way I wanted it and so on, to be put in the B-24s."[6]

Bowen meanwhile had decided that he wanted to return to Britain. Robinson's task would be to replace him. It was a tall order, particularly since Bowen had, right from the start, formed a strong and productive bond with Loomis, the man who occupied that pivotal role connecting the laboratory with the sources of political power and money in Washington. The relationship between Loomis and Robinson would be vital to the deepening technological links between Britain and the United States. Fortunately they hit it off. Robinson found Loomis "very friendly and very, very accommodating," though he clearly felt he had a lot to live up to if he were to step out from the shadow of the Welshman. "He expected me to know as much as Bowen, which was a disappointment to him," said the self-effacing Englishman.[7] In the event, the two British scientists would criss-cross the Atlantic on a regular basis in the years to come, complementing each other and providing a vital two-way flow of information about the latest techniques and developments.

By the summer of 1941, with Russia now joined as an ally against the Nazis, with AI and ASV radar making their presence felt in the night skies over Britain and in the hunt for U-boats in the Atlantic, the tide of war was beginning to turn. Microwave versions of AI and ASV had started to appear on both sides of the Atlantic, and a healthy competition had been engendered between Rad Lab and TRE. The Tizard mission's Project 1 had expanded from airborne interception to include air to surface vessel systems as well, and successful development was taking place on both fronts simultaneously. Project 2, gun-laying radar, had taken off just as well. The SCR-584 promised to be even more sophisticated than had originally been envisaged. Project 3, the search for a long-range navigation system, was coming along nicely in the background. Now, with Germany occupied on the Eastern Front, and with the balance shifting subtly in the west, thoughts were beginning to turn from how radar and microwaves might be used for defensive purposes, to how they might be corralled into assisting in the RAF's own bombing offensive.

While Tizard had been heavily involved in establishing a technological relationship with the U.S. through the mission, Frederick Lindemann had been busy consolidating his now unassailable position as Churchill's scientific right-hand man. In the late summer of 1941 he commissioned a report (the Butt Report, named for Lindemann's assistant David Benusson-Butt, who conducted the research) aimed at assessing the accuracy and efficacy of Britain's bombers operating over occupied Europe. Until now the word of the pilots themselves was the only way of knowing the accuracy or effectiveness a bombing run. Lindemann decided this was not good enough, and had bombers fitted with reconnaissance cameras mounted underneath, triggered by the bomb release. Well over 600 photographs were taken using this system, which were then compared to the accounts given by the pilots. The results, first circulated in August 1941, were shocking. The photographs, taken during night bombing raids in June and July, showed that of those aircraft recorded as having found and attacked their target, only one in three actually got within five miles of the correct target. Over the industrial Ruhr the situation was far worse, and because the report did not include those aircraft that dropped no bombs because of equipment failure, enemy action, bad weather, or just getting lost, the reality was that only about 5 percent of bombers setting out for Europe succeeded in dropping their bombs *even within five miles of their target.* If the bombing offensive was to play a major part in winning the war, then something drastic would have to be done to allow the bombers to know exactly where they were and to find and recognize their targets. In the wake of the Butt Report's findings, systematic bombing of

Germany was virtually halted. It would not resume until the adoption of area bombing tactics in February 1942.

The British already had their short-range GEE system, in which three radio stations, spaced some miles apart, transmitted synchronized signals to airplanes to enable the on-board navigator to plot his position on an overlay map. It was already showing that it could deliver hitherto impossible accuracy in navigation. But its range was limited to just a few hundred miles. Hence Project 3 on the Tizard mission's wish list—a long-range navigation system. The system envisaged by Loomis, which had been developed over the previous six months by a Rad Lab project team led by Jack Pierce from Harvard, was a long-wave system that required intersecting signals from two transmitters as much as a thousand miles apart. Synchronizing the two transmitting stations using a pair of quartz clocks would allow the target plane to calculate its exact position by means of the time difference between the receipt of signals dispatched from the two stations at the same moment. The use of a long wavelength (a key difference from GEE) meant that the signals could be bent over the horizon by bouncing them off the ionosphere. The system was originally called LRN, but later came to be known as LORAN, for LOng RAnge Navigation. It was the only major Rad Lab product *not* to be based on microwaves and the cavity magnetron.

LORAN was to play a significant part in the coming war; but for the moment the battle remained relatively short range, as Britain's bombers set out to attack targets in Germany and occupied Europe. By the fall of 1941 a solution had been found using the cavity magnetron. It brought with it a major controversy. Once the deeply disturbing findings of the Butt Report had sunk in along with the full magnitude of the inadequacy of Britain's bombers, Lindemann turned to TRE for an answer. Rowe recalled:

> The nature of Bomber Command's problem was clear. There could be no dependence upon the human eye for locating targets at night and the more aircraft used in a raid the greater the waste of effort. Somehow or other, aircraft had to reach and bomb their targets without relying upon any ground object being seen. Taking stock, it was seen that at

> least we had made a start. GEE was coming as a navigational device, though its accuracy of position-finding did not suffice for blind bombing. . . . It was [Lindemann] in particular who wanted a device with an unlimited range of operation; his eyes were already set on Berlin.[8]

Toward the end of October 1941, Rowe held a "Sunday Soviet" brainstorming session to explore how the scientists at TRE could help Bomber Command to bomb accurately targets that they could not even see—a system of blind navigation. The focus on Berlin ruled out the use of GEE or OBOE, another TRE system that was being developed with a view to attacking the factories in the Ruhr valley. Beyond three hundred miles the position of the plane could be established only to somewhere in an elliptical area six miles by one. Deeper penetration into Germany would require something fundamentally different. Discussion at the meeting quickly homed in on the idea of putting equipment into the airplane itself (GEE and OBOE both worked by transmitted signals from a home base to the airplane). Perhaps some self-sufficient equipment might be devised to follow the path of electrical power lines, or to detect cities by the magnetic field generated by their electrical installations? In his memoirs Rowe recalled that "the day ended sadly," because no workable solution was found. But two of the team working on new applications of the cavity magnetron and centimetric wavelengths soon came up with a bright idea. Philip Dee and Herbert Skinner continued to talk about the problem and recalled that early testing of ASV systems at Swanage had produced radar echoes from the town itself. Centimetric ASV systems were already showing that they could clearly differentiate between a ship and the surrounding sea. The hard surfaces of the vessel produced a much stronger echo than the rolling softness of the water. So perhaps the hard, flat surfaces of a town might stand out from the softer, more diffuse, echoes from trees, hills, and grassland.

Why not, then, let a rotating centimeter beam scan the surrounding countryside to produce a picture of towns instead of ships? A centimeter airborne intercept set at Christchurch aerodrome was quickly modified to provide a beam which could scan the ground area ahead of the aircraft, and on the Saturday following the Sunday

Soviet a flight was made to Southampton. Strong indications of the town were seen on the cathode-ray screen. Thus was blind navigation, later to be enigmatically named H2S, born.[9]

Bernard Lovell, later knighted for his work on radio astronomy, was one of the core team of scientists who was assigned to the development of blind navigation following this meeting. He later recalled the details of that flight, which would soon change the nature of the bombing campaign.

> At the time of Cherwell's [Lindemann was later ennobled, taking the title Lord Cherwell] visit in October 1941 a 10 cm AI system in a Blenheim was being tested by using a rotating paraboloid in the nose of the aircraft. Dee asked B. J. O'Kane and G. S. Hensby, who were carrying out these tests, to modify the scanner to rotate at a depressed elevation so that echoes from the ground over which the aircraft was flying could be observed. O'Kane and Hensby first flew the system at a height of 5,000 ft, and echoes from the town of Salisbury were clearly visible on the cathode-ray tube. After several more encouraging flights they photographed the cathode-ray tube: the photographs when flying over Salisbury Plain clearly revealed the towns of Salisbury and Warminster and military encampments.[10]

There are conflicting stories of its origins, but the name H2S was reportedly chosen by Lindemann as an adaptation of the familiar phrase "Home Sweet Home" because the system would allow planes to "home in" on their target. It was an ability that was soon to be demonstrated in dramatic style when test flights were made above cloud cover. Though nothing of the towns below was visible with the naked eye, the radar screen clearly showed their outline. Photographs of the screen images were taken to Rowe back at Malvern after the flight, and when he examined the prints, still wet from the darkroom, he was heard to say "This is the turning point of the war."[11] It would certainly help to turn the tide, but it would be more than a year before H2S would make its mark.

On the morning of August 9, the British battleship HMS *Prince of Wales* sailed into Placentia Bay, Newfoundland, past a line of U.S. ships to berth alongside the USS *Augusta*. On board the *Prince of Wales* was the British prime minister, Winston Churchill. His journey had been begun in secret with a cover story disseminated to the press that he was attending a flag day. The American public had been similarly duped, having been told that President Roosevelt was taking a ten-day fishing break in New England. In fact he was waiting for Churchill on the *Augusta*. The two long-time correspondents were about to open face-to-face discussions.

Unfortunately the men had rather different agendas. Churchill had come in the belief that the U.S. was on the point of entering the war against Germany, confident that he could wring such a commitment out of the president. Roosevelt, meanwhile, wanted to contain British imperialism. Now that Russia had been brought into the war he was keen to make sure that Britain and Russia came to no agreements about carving up postwar Europe. He wanted to nail down Britain's side of the bargain on Lend-Lease—a commitment to a liberalized trade regime and an end to imperial preference. It was an inauspicious start to a meeting that would lay much of the groundwork for the postwar order.

Churchill was bitterly disappointed that America was still not prepared to enter the fray. In effect he'd been sandbagged. Falling out with the Americans (and bringing Lend-Lease to a shuddering halt) was simply not an option, so when Roosevelt suggested that the conference should produce a statement of war aims, he had nowhere to go. The best Churchill could hope for was to modify the terms of the policy sought by Roosevelt and his foreign policy advisers. The document the two men eventually agreed to would come to be known as the Atlantic Charter. It was at once a prescription for international relations beyond the European war, and the beginnings of the eclipse of British power on the international stage. If there was to be a democratic future it would have a distinctly American flavor.

Roosevelt's failure to commit his country to war was a grave disappointment to Churchill. But events on the Atlantic seaways early the next month would change this. On September 4, the USS *Greer*, an American destroyer, was attacked by a German U-boat off the

Icelandic coast. The *Greer* pursued the submarine for several hours, laying down a pattern of depth charges before breaking off the engagement. Americans were incensed at this apparent German aggression toward their neutral nation. A week later Roosevelt capitalized on the public outrage by announcing a new policy in the Atlantic that would come to be known as "shoot on sight." Germany, he said, had been guilty of an act of piracy. Henceforth American ships and planes were to take offensive action "in the waters which we deem necessary for our defense. American naval vessels and American planes will no longer wait until Axis submarines lurking under the water, or Axis raiders on the surface of the sea, strike their deadly blow first." It was, in effect, the opening of hostilities in the Atlantic even though he stressed that the United States's concern "is solely defense. . . . There will be no shooting unless Germany continues to seek it."[12]

The truth of the engagement was a little murkier than Roosevelt chose to portray it. It came just four days after the announcement in Washington of a new policy in the Atlantic. The U.S. Atlantic Fleet was to begin a Denmark Strait Patrol in the stretch of water separating Iceland and Greenland. Two heavy cruisers and four destroyers had been allocated the task of escorting Atlantic convoys containing American merchant vessels—convoys that would be obvious targets for German U-boats. Implicit in the commencement of convoy duties, of course, was the requirement to take on potential attackers. What Roosevelt did not tell his listeners about the attack on the *Greer* was that for several hours before the hostilities began the destroyer had been actively assisting a British bomber in tracking the submarine U-652. The RAF plane had already dropped depth charges on the U-boat before flying off to refuel and rearm. Meanwhile the *Greer* was attempting to monitor the U-boat's position until the bomber returned. Roosevelt was deliberately vague about these facts when he spoke to the nation on September 11, 1941, "I tell you the blunt fact that the German submarine fired first upon this American destroyer without warning, and with the deliberate design to sink her."[13]

The *Greer* incident provided the pretext Roosevelt sought to move the United States closer to outright war. The powers vested in

Roosevelt by the Lend-Lease Act allowed the president to determine which countries were essential to the defense of the U.S., and now that Russia was joined with Britain in the fight against Germany he wanted to extend material support to Russia too. But he was still bound by the provisions of the 1939 Neutrality Act, and he judged that there would be substantial public opposition to any plan to repeal it. Roosevelt's answer was typical—chip away at the legislation one clause at a time. With U.S. Navy ships already on active escort duty, it seemed logical to attack Article Six, banning the arming of merchant vessels, first. On October 9, the president told the House that there was nothing belligerent about repealing this provision, that it was simply "a matter of essential defense of American rights."[14] On the night before the House was due to vote a U.S. destroyer, the *Kearny*, was torpedoed by a U-boat off Iceland. Eleven men died. The House approved the repeal of Article Six by 259 votes to 138.

Now Roosevelt and his aides stepped up the campaign. When the repeal of Article Six came up before the Senate, Democratic managers added two more provisions, this time removing the prohibition on U.S. vessels entering belligerent ports and allowing the president to declare combat zones around belligerent countries. Once again German U-boats came to the assistance. On October 31, the U.S. destroyer *Reuben James* was sunk in the North Atlantic with even greater loss of life. One hundred and fifteen men went down with the *Reuben James*. A week later, on November 7, the Senate voted through all provisions, effectively removing the main constraints imposed on the commander-in-chief by the Neutrality Act. Bit by bit America was preparing for war.

To Roosevelt, Germany was the principal danger to the world order; and with war already being waged in Europe, he had taken personal charge of that sphere of U.S. foreign policy. Japan, by contrast, was thought of as something of a hollow threat, whose ambitions could be placated and met through concerted diplomacy. So while the president concentrated on events in the North Atlantic and Europe, Far Eastern affairs were left to Secretary of State Cordell Hull.

Two principal issues divided Japan and America in the fall of 1941, China and the Tripartite Pact. America and Britain both supported

China in the defensive war against Japan she had been waging for several years, and American policy demanded that Japan withdraw its million troops from China. Japan's membership of the Tripartite Pact set it firmly on the side of those in pursuit of a new Fascist world order which ran directly counter to Hull's Wilsonian principles of free trade and self-determination. Just as in America, opinion in Japan was split. The government of Prime Minister Fumimaro Konoe was keen to reach an accommodation with America so as to restore access to trading, and in particular to American oil, supplies of which had been suspended in July. The Japanese Army, however, was determined to resist even token withdrawals of troops from China, seeing it as a return to the old days when Western imperial powers had controlled events in Asia.

In early October the State Department withdrew from planned talks after the Japanese had been unable to agree to a statement of shared principles on sovereignty and noninterference given to them some months earlier by the Americans. American withdrawal from the talks was a body blow to the government of Prime Minister Konoe. On October 16, his government fell. The new prime minister would be the Army Minister, General Tojo Hideki—the military was strengthening its grip on Japan. The first two days of November saw a joint liaison conference of Japan's Army and Navy. The internal political momentum of Japan was undoubtedly heading towards war, but they decided to give the new Foreign Minister, Togo Shigenori, until the end of the month to reach agreement with the U.S. Their new plan came in two parts, the first being the uncompromising setting out of Japan's own terms for a settlement with the U.S. These, they knew, would be unacceptable, so a fallback position was also prepared that would call for America to lift its oil embargo in return for Japan's withdrawing its troops from Indo-China. No offer was made to withdraw from China.

Accepting "no withdrawal" was not an option for Hull, raising, as it did, the awful specter of the appeasement that had brought on the disastrous situation in Europe. Chiang Kai-shek warned that the Chinese people would see it as a complete betrayal. Churchill waded in in support of the Chinese. A Chinese collapse was a distinct possibility unless the U.S. stood firm, and a Chinese collapse would have dire consequences for the Russians, who would then find themselves fighting on two fronts.

Since 1939 the Americans had been at a distinct advantage in their dealings with Japan, courtesy of their cryptographers. The Japanese Foreign Service used coding/decoding machines to encrypt their signals traffic—a system known to the British and Americans as Purple. In 1940 the Army Systems Intelligence Service (SIS) cracked the Purple code using a technique that came to be known as MAGIC. By the tail end of 1941 these decrypts were of limited use, as power in Japan had shifted towards the military, who used a different cryptographic system. But through MAGIC intercepts, Secretary of State Hull was aware of the Japanese fallback position well before it was put, and by November 25, he knew, too, that a large Japanese force had been embarked at Shanghai and was en route to Indo-China and the Malay Peninsula. It was clear that Japan had no intention of negotiating a way forward acceptable to the Americans. From the decrypts the State Department knew that Japan's new military-dominated government had put a deadline of late November on the attempt to reach a negotiated agreement. Hull, whose policy in the East had been marked by a desire to avoid war, was now forced to accept that war was almost inevitable. On November 26, Hull handed to the Japanese a ten-point plan he knew would be unacceptable. The following day he told Stimson and Knox that the State Department had reached the end of the line—the situation in the Pacific would henceforth be their province.

But if war was to come, who would strike the first blow? And where? The U.S. policy of strengthening air defenses in the Philippines was far from complete, as the new Boeing B-17s allocated to the job had yet to arrive. America still wanted to buy time to complete its build-up of troops and air defenses in the Far East. Large troop movements were scheduled for November and December, but they would not be complete until the early spring. Climatic conditions, however, implied that any Japanese aggression in Southeast Asia would come sooner rather than later—before the arrival of the northeasterly monsoon at the turn of the year. That would suit Roosevelt, who, as in the Atlantic, was determined that the U.S. should go to war in response to provocation, rather than be painted as an aggressor.

In fact as the calendar moved into the momentous month that would bring war to America, governments on both sides of the Atlantic were convinced that when war came it would most likely be through an attack, not on American interests, but on British or Dutch territory in South East Asia—probably in Malaya or the Dutch East Indies. With America having declined British requests to move its Pacific Fleet further west than Hawaii, the key question was how much support the U.S. would offer if the Japanese attacked British interests. On December 3, Lord Halifax, the new British Ambassador, pressed Roosevelt to come down from the fence. Roosevelt told Halifax he had no doubt that in those circumstances "you can count on [the] armed support of [the] United States."[15] He would have to persuade a reluctant Congress to go to war in defense of British imperial interests. Instead, war came to America.

The attack took place without any formal declaration of war by Japan. In a token attempt to uphold the conventions of war while still achieving surprise, it seems the Japanese intended that the attack should begin some thirty minutes after its ambassador had informed the United States that peace negotiations were at an end (even though the message Nomura was asked to deliver contained no actual declaration of war). In the event though, the attack began before the five-thousand-word notice could be transcribed and delivered by the embassy. On this particular Sunday, information that could—at the very least—have raised the state of preparedness at Pearl Harbor simply did not get through, either from the Japanese or from American decrypts of that same message from Tokyo.

There is a world of difference between a country at war and one having war thrust upon it. So pervasive was the conviction that any Japanese attack would come in the Far East, so solid the certainty that the Japanese could not strike across thousands of miles of ocean, and so lamentable the intelligence on the movements of its navy, that even on the brink of war Pearl Harbor lay shrouded in a cloak of complacency the morning of December 7. That lack of preparedness denied the Navy warning they could have had from radar. Three months before the attack five mobile radar sets, U.S. Army SCR-270s, had been positioned around Oahu. They were to provide a protective screen for the Pacific Fleet extending some 150 miles out into the Pacific—once their operators were fully trained. On the morning of December 7, they had been in action since 4 A.M., still being operated in training mode. They were due to stay on alert until seven, just about an hour after the first Japanese bombers took off from their carriers some 220 miles out at sea. At 6:45 A.M. three of the five stations detected faint echoes on their screens, but in two of them the traces were disregarded, possibly on the grounds that they represented flocks of birds, and the stations were duly shut down at 7 A.M. At the third station, Opana, ideally situated some 230 feet above the sea, the inexperienced operator, Private George E. Elliott, asked for a little more time on the system with his instructor, Private Joseph L. Lockard. Suddenly the cathode ray tube was filled with so many echoes they thought the system must be malfunctioning. A quick system check confirmed the SCR-270 to be operating properly, and the two men were left with the terrifying conclusion that a large aircraft formation was approaching Oahu. At Elliott's suggestion, Lockard tried to telephone the findings through to the Intercept Center, where all their reconnaissance reports were collated. It was now just past seven o'clock and Lieutenant Kermit Tyler was the only man left on duty. Tyler himself was still in training, and the Intercept Center only partially activated. Tyler knew that a flight of B-17s was expected to arrive at Pearl Harbor shortly—information the radar operators were unaware of, and that he was not empowered to tell them. The radar traces seemed to show aircraft coming from where the B-17s were expected; so surely these echoes must be those of the arriving USAAF planes? He remembered, too, that the Army had an arrangement with a local radio station to play music throughout the night whenever planes were due, in order to give them a radio signal to home in on. On his way to the center in the early hours of the morning he had listened to music being played on the station. It was enough to convince him that the echoes reported by Elliott and Lockard must be the incoming B-17s. He told the two men to ignore the radar signals, uttering the fateful, and now famous, words: "Don't worry about it."[16] Elliott and Lockard had done their duty. They sat and watched the screen as it plotted the progress of almost two hundred Japanese bombers carrying the torpedoes and bombs that would soon sink four battleships, destroy or damage almost 350 aircraft, and take the lives of 2,335 serving men and 68 civilians.

That the Japanese would strike directly at America was unexpected. That they should have laid waste the Pacific Fleet came as a profound shock to Roosevelt. The event was at the same time a vindication of his view that the coming war would be dominated by airpower, and a devastating blow to his belief in the power of the U.S. Navy. Frances Perkins, one member of the Cabinet that Roosevelt gathered around him that afternoon, recalled that: "he was having actual physical difficulty in getting out the words that put him on record as knowing that the Navy was caught unaware, that bombs dropped on ships that were not in fighting shape and not prepared to move, but were just tied up."[17]

The following day Roosevelt went before Congress to tell them that a state of war had existed between America and Japan "since the unprovoked and dastardly attack by Japan." Britain followed, and in quick succession, Australia, New Zealand, the Netherlands, the Free French, and several South American countries. China declared war on Germany and Italy, and formalized the state of undeclared war that had been escalating between itself and Japan since the Japanese invasion of Manchuria a decade earlier. The Japanese, for their part, attacked the Philippines with a massive air bombardment that destroyed most of the U.S. air defenses on the ground. British territory, too, came under attack with Japanese landings in Malaya and southern Thailand. Most of the RAF's 158 airplanes in Malaya were destroyed on the ground, and the airfield at Kota Bharu was captured intact by the Japanese. Hong Kong, Britain's ancient trading post on the South China coast, came under heavy assault, forcing an ignominious retreat by the British garrison. In China, Japanese troops occupied Shanghai and captured a small U.S. garrison in the foreign section. By the close of that extraordinary day the fiction of Japan as a hollow threat had been firmly laid to rest.

On December 9, Roosevelt spoke to the nation in his most important fireside chat to date. Not only did he tell them of events in the Pacific, but he warned them of worse to come—that the opening of war in the west was directly connected to the situation in Europe: "Your government knows that for weeks Germany has been telling Japan that if Japan did not attack the United States, Japan would not share in dividing the spoils with Germany when the peace came. . . . We know also that Germany and Japan are conducting their military and naval operations in accordance with a joint plan."[18]

He spoke of the grand strategy of the Axis, and said that the only way to oppose it was "with a similar Grand Strategy" directed as much at Germany as Japan. It was a prescription for the way this new war would unfold for the United States, and it came to pass sooner than he might have expected when Hitler and Mussolini declared war on the United States two days later, on December 11.

This fireside chat concealed one truth and engendered another. The grand strategy of the Axis powers was in fact something of a hollow alliance, as the three powers were in no way committed to joint aggressive action. Hitler had encouraged Japan to go to war with the U.S. because he wanted America occupied away from Europe. His declaration of war against the United States was his decision alone, and the two powers went on to fight separate wars against the newly allied Britain and United States. Conversely, the new community of interest between Britain and America would grow into a solid alliance combining the strengths of the outgoing imperial power of Britain with the rising industrial, military, and economic might of America to devastating effect. It was a partnership that had been sown with the correspondence between Churchill and Roosevelt, nurtured by the scientific exchanges of the Tizard mission, and was now about to flower on the field of battle with new weapons and systems that would change the face of the war America had now, at last, entered. Radar had peered through the ether to find the incoming raiders that morning of December 7, but its lessons had gone unheeded. In the months that followed Pearl Harbor those lessons would be watched and listened to with ever greater care as the United States went to battle with the new generation of radar—powered by microwaves and developed from the contents of Sir Henry Tizard's black japanned deed box.

SEVENTEEN

Secret Weapons

By December 1941 the U.S. military had understood the importance of radar (even if Pearl Harbor showed there was much still to be learned about its application); but when this slumbering giant finally went to war, as the Office of Scientific Research and Design's own report published at the end of the war showed, microwave technologies remained just beyond the horizon:

> By the time of the Japanese attack on Pearl Harbor the Navy had already installed on key ships not only radar for air warning, but also medium-wave radar for surface search and fire control, while the Army had deployed in the field numbers of long-range aircraft warning sets, as well as anti-aircraft gun and searchlight batteries equipped with radar. Navy installation of equipment on more and more ships, and Army training of additional Signal Aircraft Warning Battalions and anti-aircraft batteries were proceeding at full speed by the end of 1941. During that year, by contrast, not a single item of equipment based on the new microwave development was delivered for operational use.[1]

During November 1941, with war on the horizon, the NDRC established a model shop at Rad Lab aiming to transform the customized creations of the scientists into something that could be produced on an industrial scale and used, not by elite academic physicists, but now by young servicemen, often in the most arduous conditions and while their lives were under threat. Bridging that gap would be the work of 1942.

By the summer of 1942 new U-boats were coming from Germany's shipyards faster than the Allies were able to destroy them. They had been winning the battle—particularly in the vital arena of the Bay of Biscay and the southern approaches to the Atlantic. After America's entry into the war things deteriorated further as the submarines shifted their attentions toward relatively unprotected U.S. shipping off the East Coast of America. But in June 1942, a new weapon appeared that would fill the breach until the arrival of microwave systems. From the beginning of the war U-boats had been almost completely immune from attack at night. They had taken advantage of this weakness, remaining submerged (and invisible to radar sweeps) during the day, then surfacing at night to recharge their batteries under cover of darkness. But early in June, everything changed with the appearance over the Bay of Biscay of Wellington bombers equipped with metric radar (Bowen's ASV MkII) and a powerful searchlight called a Leigh Light, which at a stroke released the potential of radar at night. Until now a patroling aircraft detecting the presence of a submarine on the surface would lose sight of it on the final run in, as reflections from the sea cluttered their radar screen and drowned out the signal. The Leigh Light changed all that. Now the aircraft would follow the radar trace to within a mile of the target, then switch on the powerful searchlight, using the light to hold the submarine in its full glare until the aircraft was overhead and could drop its depth charges. The new system had a profound effect on the precarious balance of the battle between U-boats and the air patrols. From December 1940 through February 1941, 96 Allied ships had been lost, with not a single U-boat sunk. After the arrival of ASV MkII the situation improved, and with the introduction of the Leigh Light, the balance swung dramatically in favor of the aircraft.

The Kriegsmarine responded with a change of tactics. Now the Germans took the view that their U-boats stood a better chance of seeing approaching aircraft during daylight, so they began to recharge their batteries in daytime rather than at night. The RAF's Coastal Command responded with an increase in day patrols. The pickings were good. For a while things went well for the airborne hunters, until mysteriously, U-boats began disappearing off the radar screens of the attacking aircraft. Somehow they could now see the radar-equipped planes coming.

The explanation lay in the loss of a British bomber in the clear skies over North Africa. Following the shooting down in Tunisia of a Wellington equipped with ASV MkII, German general Erwin Rommel had been using it for reconnaissance in the Mediterranean so successfully that it was taken back to Germany so that the system could be copied. But Germany's scientists also went one step further. They managed to design a listening device, given the name Metox, that could detect the signals from an ASV MkII and give the U-boat commander plenty of time to crash dive and maneuver his craft away from the depth charges below the surface. Thus the balance shifted again—this time in favor of the submarine. It would not be until the arrival of microwave ASV in March 1943 that the Allies would once again gain the upper hand.

The story of microwave ASV had started more than a year earlier and on the other side of the Atlantic. Galvanized by Pearl Harbor, the USAAF had moved quickly to request the installation of Rad Lab's microwave radar in some of its aircraft as early as possible. Ten Douglas B-18s were to be equipped with handmade systems. By April 1942, this squadron was in the air and the new system was being used out of Langley Field in the search for submarines off the East Coast. A production contract was placed for the modification of a batch of airborne interception sets that were about to come out of the factory to be adapted for ASV use, and soon there were several hundred bombers so equipped. Their immediate success forced the U-boats to move further and further south. They were followed by what had now grown into the Army Air Forces Antisubmarine Command, working under the operational control of the Navy, who were responsible for protecting the seaways. The British, suffering as they were from the recent change in U-boat tactics, requested a field test of the equipment. In March 1942, a Lend-Lease B-24 Liberator flew to England for trials with the original laboratory prototype. Denis Robinson was returning home to demonstrate what Rad Lab had managed to achieve in the remarkably short time since he arrived with the request. It was fitted with the best new technology available on either side of the Atlantic. He was asked to demonstrate its capabilities in Northern Ireland, where a small submarine was made available on one of the big inland lakes. The results were remarkable, as Robinson later explained: "we got results, at all different heights, all kinds of this, that and the other. I flew back to London to see Air Marshal Joubert. . . . I think he had the responsibility for all short-wave radar at the time. I went to see him at one of the big air bases, and he said, 'Robinson, what the hell are you sitting there for? Get back across to America and get more of these!' I'd brought him the results, and they were clearly better than anything the British had then."[2]

The superiority of the system was clear and Coastal Command soon ordered production of a British version. This was the system that came into operation as ASV MkIII in February and March 1943. The ASV MkIII, a cavity magnetron at its core and equipped with a Leigh Light, was a formidable weapon. Eddie Bowen thought that its introduction to Coastal Command was "the beginning of the end for the German Submarine Service."[3] Metox receivers were useless against centimetric radar, rendering the submarines sitting ducks by day, while by night they could be illuminated and attacked with even greater accuracy than before. In the three months from May to July nearly one hundred German submarines were destroyed—two-thirds of them by aircraft.[4] The tide of the Battle of the Atlantic was making what would prove to be an irreversible turn.

The Germans' first indication that something was wrong had been when Metox stopped working. Suddenly the enemy had a radar system that could not be detected by Metox. The explanation seemed simple—ASV MkII had simply been modified in some way. Being in possession of a captured MkII they were soon able to find a modification that made it invisible to Metox. This was good news—it was a modification that could easily be countered, and so they made the necessary changes to Metox and put to sea with a test rig. In fact the change in the Allied system was, of course, the much more significant switch from metric to centimetric radar; but this was unknown to the German technicians. The intense secrecy surrounding the cavity magnetron since the very beginning of the Tizard mission was now paying off.

Yet by a strange twist of fate, in their first encounter with an Allied plane at sea with their modified Metox the Germans found themselves not against a plane carrying one of the new MkIII microwave systems, but against an old MkII that had been experimentally modified in just the way the Germans had predicted was being applied to the whole fleet of enemy aircraft. The modified Metox had run into just about the only Allied plane against which it would work. So the technicians returned to base confident that they had the answer, and began equipping their submarines accordingly. But the Allied success was not stemmed and the submarine kills continued unabated.

The Germans realized that the Allies must be using some new technology, but they decided it must be some kind of infrared ray and started to paint their submarines with a special paint. They guessed wrong, and the submarine kills went on. By now the crews were losing confidence in their technicians, and the job was handed over to scientists, both military and civilian, who equipped a U-boat with a whole array of detection equipment to try to find out what was going on. Thirteen days later it was sunk. A second expedition was quickly put together, but this one lasted just nine days. A little over a month later Allied troops landed on the beaches of northern France. U-boats quickly became irrelevant as the Allied land armies ground their way across Europe toward Berlin. The battle of the Atlantic was over.

Long before the U-boats had been finally defeated, Germany's aerial assault on the UK had lost much of its impetus. Fighter Command's stout defense and the increasing sophistication of the radar aids at its disposal, together with improved anti-aircraft gunnery and the seemingly indefatigable morale of the people, were a big part of the story. So too, though, were increasing German losses, and Hitler's decision to throw most of his resources into the new battle he had opened up in the east against Russia in the summer of 1941. Britain could now begin to contemplate a concerted bombing attack on Germany.

The Butt Report, though, had revealed Britain's bombers to be hopelessly inaccurate—risking men and machines while accomplishing little of the devastation expected of them. British targets in Germany (particularly Berlin) could be as much as 500–700 miles away from the GEE transmitters in the UK, hidden behind the curvature of the earth. This was a major problem. During 1942 one of TRE's brightest scientists, Alec Reeves, and his assistant F. E. Jones developed the blind-bombing system Oboe, which could control the position of the bomber in the skies over Europe with a greater degree of accuracy. Two transmitting stations on the English coast used pulsed radio transmissions to hold an airplane on a curved path that would take it directly over the target, then to trigger the precise moment of bomb release.

The signals from Oboe's two transmitters (referred to as Cat and Mouse) passed over the horizon at sea level then began to climb as the earth curved away beneath. By the time the signal reached 250–270 miles it had reached a height of around 30,000 feet. The use of the Mosquito, a recently introduced twin-engine fighter-bomber designed and built by de Havilland, was vital because it was both high-powered and extremely light (unusually at this time, it had a wooden airframe), which meant it could operate at heights unreachable by other aircraft big enough to take the system. Sitting in safety in England, Oboe operators could follow the track of a Mosquito aircraft and could calculate its speed. The Mosquito crew had little more to do than obey simple indications and to release their flares over the target when signaled to do so from England.[5] Oboe could only control one aircraft at a time, so Oboe-equipped Mosquitoes were introduced as the leading edge of a new Pathfinder Force; their task was to drop target-indicator flares with the highest possible degree of accuracy, so that the main bomber force following up at a much lower altitude could drop their bombs on to the precisely marked targets. The introduction of Oboe-equipped Pathfinders would prove devastating to the German industrial heartland of the Ruhr. As A. P. Rowe later recalled:

> Rarely in war is the first use of a device attended with the brilliant success achieved by Oboe on 21 December 1942. Hitherto all the efforts of the aircraft firms to produce heavy bombers, all the brilliant organization of Bomber Command and all the courage of the bomber crews had failed to damage Krupp's Works near Essen. But on the night of 21 December it was estimated that 50 percent of the bombs fell on the target. This was but a beginning. . . . Oboe, first with a 1.5 m. wavelength and then with 10 cm. devastated the Ruhr and other areas.[6]

If Allied bombers were to be able to hit Berlin itself, however, at some three hundred miles further than the Ruhr, a different approach would be needed, something that could travel with the aircraft itself, maybe. Here, of course, the cavity magnetron would be key. At the close of October 1941, Rowe's special Sunday Soviet at TRE had set Britain's radar scientists on the path that quickly produced H2S, a new implementation of microwave radar, adapted from ASV, which, when installed in an aircraft along with a cathode ray plan position indicator, presented its crew with a radar picture of the terrain they were overflying. Experience during the Battle of Britain and the Blitz, and their own bombers' bloody exposure in early raids on Germany, had convinced the British to abandon daytime bombing and to copy the German tactics of nighttime attacks against which it had proved so difficult to defend. If accurate navigation was hard during the daytime, at night the obstacles were almost insurmountable without technical assistance.

There was a difficulty. So far the available navigational aids such as GEE and Oboe had their key equipment safely positioned in static facilities on British soil. Having similar equipment on a plane was now made possible by the cavity magnetron—but that device was a crucial technological advantage held by the Allies. What if a plane carrying H2S and the magnetron were shot down over Germany? Wouldn't this hand them microwave radar on a plate? A system was designed to try to destroy the H2S equipment in the event of a plane coming down behind enemy lines, but the magnetron itself proved to be virtually indestructible. Using it over the sea, on ASV missions for instance, was not a problem—a downed plane would simply sink, taking the magnetron with it. Crash-landing in German-controlled territory was quite another matter.

So significant was the fear of the cavity magnetron's falling into German hands that first attempts to design and build a production version of H2S tried to use a klystron as the power source. The contract was given to EMI (Electrical and Musical Industries). This company had invaluable experience in the design of emerging television technologies, which suited them well to a project of which a cathode-ray display was a crucial part. In their chief development scientist, Alan Blumlein, they had a genius of circuit design. Their task was however made yet more difficult by the requirement to install H2S in the new range of four-engine bombers flying at 20,000 feet (about three times as high as the test flight). Progress was by no means fast enough for the impatient Churchill. On June 7, 1942, he wrote to the Secretary of State for Air.

> I have learnt with pleasure that the preliminary trials of H2S have been extremely satisfactory. But I am deeply disturbed at the very slow rate of progress promised for its production. Three sets in August and twelve in November is not even beginning to touch the problem. We must insist on getting, at any rate, a sufficient number to light up the target by the autumn, even if we cannot get them into all the bombers, and nothing should be allowed to stand in the way of this.[7]

By a sad irony his memorandum was to be overtaken when, on that same day, a Halifax bomber carrying the latest (and only functioning) H2S set on a test flight crashed in the Wye Valley, killing all eleven men on board, including Alan Blumlein and his team. It was a major setback, but it would also play a big part in convincing the politicians that there was no alternative to putting the cavity magnetron at risk.

When Churchill called another meeting on July 3, summoning Lovell among others, he insisted that the crash of June 7 should not be allowed to hold up H2S. Lovell observed: "He showed little sympathy with the loss of the Halifax carrying the only working H2S equipment. He said that was only one aircraft and we had lost 30 over the Ruhr in the previous night. When told of the difficulties he thumped the table and said he must have two squadrons fitted by October: 'Our only means of holding the enemy.'"[8]

And on July 15 came the news that the Secretary of State for Air had ruled that development work on the klystron for H2S should cease. Churchill's two H2S squadrons would be equipped with a magnetron version. There was to be a rush program for the production of H2S sets, enough initially to be installed in the new Pathfinder Force, then expanding to the volume production of enough sets to equip the main bomber force. For Rowe and his men at TRE this was the heaviest demand yet:

> We had had crash programmes before but never one like this. A "crash" programme implied that instead of undertaking the normal process of tooling for production so that the minimum of highly skilled labour was used, sets should be almost handmade, whatever the demands on skilled men. The same process had to be applied to the major modifications of the bomber aircraft needed to provide room for the H2S equipment. A crash programme meant also that the design of the equipment was influenced more by the need for rapid production than by the attainment of the best performance.[9]

Once more, faced with the urgent need to take the offensive, Britain's radar scientists found themselves rigging customized systems into airplanes not designed for the task.

But the stage was now set for a major transatlantic disagreement. British planes would use GEE to find their way to the target area, and H2S to home in on the target itself. The British seemed determined to risk the precious cavity magnetron in a navigational system that, as far as the Americans could see, simply didn't work. It looked to them like an act of consummate folly. Lovell recalled:

> The belief that the magnetron was a vital Allied secret led to an extraordinary situation in July 1942. The American scientists had been informed of the development of the blind bombing H2S system, and shortly after the Prime Minister's meeting at which he demanded "two squadrons by October" an American deputation arrived in the TRE. Headed by I. I.

> Rabi, the deputy director of the Radiation Laboratory, and including E. M. Purcell, there was an unpleasant encounter at which they argued that if we persisted in using the system operationally then the Germans would quickly discover the magnetron secret and thereby possess information that would severely endanger Allied operations.[10]

The decision taken by the Secretary of State for Air in mid-July to ditch the klystron in favor of the magnetron as a power source for H2S had one key proviso. This vital technology was only to be risked in flights over Germany "if the Russians hold the line of the Volga"—meaning that the Russians successfully resisted the German onslaught in the east. In any event, for the time being the issue was academic, as Churchill's demand for two squadrons to be equipped by October simply could not be met after the loss of the EMI scientists and their prototype in the Halifax crash of June 7. Such was the priority attached to this project, and the political pressure to deliver that went with it, that barely two months after Churchill's deadline, by December 1942, two squadrons of the Pathfinder Force had each been equipped with twelve bombers fitted with H2S.

Now, for the first time, there was a possibility that Allied bombers (or at least British bombers) could navigate their way across six hundred miles of hostile territory to Berlin—and with hitherto unheard-of accuracy. But for now their hand was stayed. Events turned on the Battle of Stalingrad. After five months of fighting, late in November 1942, the Russians launched a counter-attack codenamed Operation Uranus. The Axis forces surrounding Stalingrad were taken by surprise, and by November 23, the Red Army had trapped about 220,000 German soldiers in a pocket some thirty-five miles wide and twenty miles north to south. In mid-December a German relief effort, Operation Winter Storm, managed to advance to within thirty miles of the city before it was halted on December 21. By Christmas it was clear the Russians were now in the ascendant against Germany. This was the moment Britain's Secretary of State had been waiting for. Now, he concluded, the line of the Volga would indeed be held. H2S could be prepared to fly.

On the last night of January 1943, Pathfinder Force used H2S-equipped bombers to mark targets in the city of Hamburg. At last, and somewhat against the wishes of its American allies, H2S, perhaps the most revolutionary application of the new technology the Tizard mission had carried across the Atlantic, had been thrown into battle over Europe by the British. Days later American fears were realized when an RAF Stirling equipped with H2S was shot down near Rotterdam on the night of February 2-3, 1943. It crashed just outside the city, and within a few hours the Germans had their hands on H2S and the cavity magnetron. The radar equipment was recovered almost entirely intact by radar engineers from the Telefunken company.

Professor Sir Bernard Lovell was one of the core team that developed H2S. In 1977, he was invited to visit the radio telescope at Effelsberg, near Bonn, to discuss research matters and the collaboration between radio astronomers. At dinner, after the visit was over, the director, Professor Otto Hachenberg, raised the subject of scientists in the war years and told Lovell: "I am well aware of your wartime occupation because as a young man then working in Telefunken I was sent to investigate the equipment in a bomber that crashed near Rotterdam in 1943." It was the beginning of a chain of events that would eventually, after Hachenberg's death in 2001, hand to Lovell a copy of his report on the radar equipment recovered from the Stirling, *Englisches Radar-Gerat* (English Radar Equipment). It made chastening reading.

After the Stirling bomber crashed and its wreckage recovered, a committee was convened to analyze the new equipment that had fallen so fortuitously into German hands. Called the Rotterdam Committee, it was comprised of scientists, the military, and representatives of the electronics industry. Hachenberg's company, Telefunken, had the task of deconstructing the equipment. The committee first came together to begin to discuss what could be learned from the dismantled kit at the Telefunken offices just two weeks after the Stirling came down. On May 1, Hachenberg published his report, which included further analysis of a second device that seemed to have been recovered from an H2S-equipped Halifax from Pathfinder Force that had been shot down in the interim. When Lovell sat down to read the report, some sixty years after it was written, he was amazed by what he saw.

> The report makes it clear that the Germans quickly identified the H2S as a blind bombing system, but it is the detail in the report that is so surprising. The photographs accompanying the report are of individual units, all of which are easily identified except the transmitter unit, which carried an explosive device intended to hinder identification of the cavity magnetron. In fact, the report is an accurate account of the function of every unit. . . . The great surprise in that report is . . . that the cavity magnetron is described precisely, ending with the comment that "*Der Wirkungsgrad der Magnetronanordnung, die uebrigens den Nachbau eines hier bekannten russischen Patentes darstellt, betraegt etwa 10%*" ("The efficiency of the magnetron—it is worth mentioning that this is a replica of a known Russian patent—is about 10%").[11]

Two things emerge from this report that throw a curious new light on the story of the cavity magnetron. First, and as Rabi had predicted, the Germans had had no difficulty in interpreting the equipment for what it was. In fact they immediately set about copying it, and the LMS10, a German version of the captured H2S, had been produced within three weeks. The first production order, for six sets, was quickly placed with the Telefunken company, followed by a further twenty for manufacture between June and September 1943. Second, it was clear from the report that the Germans were well aware that Russian scientists had already patented a version of the cavity magnetron—a fact that had been unknown to the British and Americans. Randall and Boot had not been alone in unlocking the power of microwaves. What was different about the experience of the Allies, however, was that they had succeeded in capitalizing on the discovery and turning it into a weapon of war in such a short time. But now, as some had feared, all this development work had fallen into the hands of the Germans. It could have been disastrous. It was not.

The Luftwaffe did put their version of H2S to use against the Allies, but to a degree dwarfed by the raids now being launched against their own cities. Oboe- and H2S-guided bombing was taking a terrible toll of German manufacturing. With H2S successfully decoded, the Rotterdam Committee looked around for companies capable of taking on its manufacture. Orders were placed with

Telefunken, and Sanitas, a company specializing in X-ray equipment. They were to manufacture three sets each per week. It was a mere drop in the ocean compared with the Allied production pouring out of factories on both sides of the Atlantic. And then things got worse when the Sanitas factory itself fell victim to an Allied raid. Nothing was coming through in any quantity until the fall of 1943—and by then Allied planes were being fitted with an even more advanced system.

German factories were not, though, the only source of H2S equipment. Again, Rabi's worries were vindicated when German planes began appearing over Britain using reconditioned British or American equipment recovered from bombers shot down over continental Europe. The heavy German attacks on London in February and March 1944, which came to be known as the "Little Blitz," were largely guided by H2S sets being turned against their makers. But by then the damage to Germany's own productive capacity was severe. Within days of its first use over Europe Germany had captured an H2S, within a few weeks they had copied it, and within months they had it in limited production along with vital defensive systems that could pick up H2S signals from incoming planes. Lovell questions the decision to delay its use, arguing that it could have saved many lives over the Bay of Biscay in the fall and winter of 1942. Nothing, he says, was gained by delaying, because the Germans were unable to put it to effective use once they got it, and in any event they already knew of the earlier Russian development of a cavity magnetron—at a time when they had had a nonaggression pact with Russia. In fact, what prevented the Germans from making better use of it once they had it was simply that they could no longer protect the factories needed to produce it in volume. That was due to the concerted bombing campaign that the RAF began in February 1942, and which was soon to be reinforced.

George Valley came to Britain in the fall of 1942. He worked on the gun-laying team at Rad Lab and had traveled across the Atlantic to exchange information and experience with British scientists working on the same problems. While passing through London he witnessed something few of his Rad Lab colleagues had ever experienced. He stood on the balcony of his hotel room in central London and watched an air raid. It was a sobering moment for him, seeing at first hand the damage bombers could wreak once they had found their target—and it convinced him that helping Allied bombers crack the problem of precise navigation would have greater impact than improving anti-aircraft gunnery. Offense had suddenly become more attractive to Valley than defense. On his return to Cambridge, Lee DuBridge agreed to transfer him to Rad Lab's NAB (Navigation and Bombing) team working with the H2S-type technology that had been passed along by the British. It was a fortuitous switch, because Valley would later solve the problems that beset American work on 10 centimeter H2S and move it to a much shorter wavelength. This improved version, operating at 3 centimeters, with commensurately greater precision and definition, would turn out to be the most successful navigational aid for the bombers that would take the campaign deep into the heart of Germany. Its official title was H2X (the wavelength it used was in the X-band instead of the S-band of H2S); but it would be known affectionately as "Mickey."

But that was all still in the future. For the moment, in the fall of 1942, the Americans had a 10 centimeter system that simply did not work. In the late summer Denis Robinson had written to Dee, back at TRE, that "H2S did not work in America."[12] Even he, an acknowledged expert, had been unable to distinguish between cities and countryside using the American system. The limiter that stopped the PPI screen flaring in response to a sudden sighting of a ship in the middle of echoes from the sea might be helpful on an ASV system, but over land it suppressed all the subtle differences that enabled the system to distinguish towns from countryside, buildings from bushes.

That summer the USAAF Eighth Air Force arrived in Britain to lend its (at first somewhat slight) weight to the Allied bombing effort. It brought with it new airplanes, fresh aircrew, and its own tactics. From the start and to the end of its operational life the Eighth was committed to daytime bombing. Many months previously the RAF had abandoned the idea of bombing during the day, adopting the enemy's own technique of nighttime raids. Hence the absolute imperative to get H2S into aircraft as soon as possible. The Eighth arrived with a different plan—precision bombing using visual bombing aids—as the Norden bombsight at last made its appearance in the war. H2S at this time did not bear that name. Officially it was known as BTO (Bombing Through Overcast), and the newly arrived Americans soon discovered why. The Norden bombsight had been developed and tested in the clear skies over the continental United States. As the fall arrived in Europe much of the continent was quickly shrouded in cloud. Visual aiming just didn't work. As winter set in the Eighth found itself virtually grounded. Hardly a plane left the airfields in December and January 1943. This was not unusual. Air Force meteorologists advised that less than half the days in every year would be suitable for visual aiming.

On February 4, 1943, the day after the RAF Stirling with its secret H2S crashed outside Rotterdam, the Allied bombing offensive took a new direction. A little over a fortnight earlier Roosevelt and Churchill had met together for the first time as allies in Casablanca. Their goal was to carve out a joint strategy, and high on their agenda was the role of the bomber forces operating out of the United Kingdom. In fact its importance became ever more apparent as the conference unfolded. Stalin, whose troops were sustaining terrible losses in bloody fighting on the eastern front, was keen that the U.S. and Britain should open up a second front in the west. Churchill, though, was adamant that it was too early to risk an invasion across the English Channel. The policy agreed was to take the war to Germany through an invasion of Sicily and southern Italy while stepping up the bomber offensive on Germany itself in an attempt to break the will of the German people. The goal of the Allies would be "the unconditional surrender of the Axis Powers." The policy agreed to at Casablanca on January 21 was issued to the respective air forces in a memorandum circulated on February 4, which stated: "Your Primary object will be the progressive destruction and dislocation of the German military, industrial, and economic system, and the undermining of the morale of the German people to a point where their capacity for armed resistance is fatally weakened."[13] To Arthur Harris, head of the RAF's Bomber Command, this was the go-ahead for a new strategy of intense area bombing. The policy set out in that report would result, its authors estimated, in the destruction of six million German homes and the deaths of as many as one million people.

On February 16, DuBridge convened a Rad Lab conference to discuss the issue of bombing aids. Valley was able to tell them that preliminary tests of his new 3 centimeter system had been promising; but even this did not seem to offer the same precision as the much vaunted Norden bombsight. Precision, until now, had been the name of the game. It was debatable, to say the least, whether the accuracy of visual bombing on the target range could ever be translated to the realities of war; but the truth of the matter now was that formation flying meant pattern bombing—and the accuracy required for pattern bombing was well within the capabilities of Valley's 3 centimeter system. Bomber Command's policy of area bombing was anathema to the USAAF; but the new practicalities of war meant accepting a move away from the ideal, blurring the edges of the target at the very least. If this was the way the war was moving, then there was surely a role for H2X. At the end of the meeting DuBridge, aware that TRE was committed to turning out hand-built H2S for the foreseeable future, told Valley that Rad Lab might perhaps develop a 3 centimeter system if the British were not going to. Valley took that as a green light. Mickey was on its way.

As the hard winter of 1942–43 came toward its end, the era of electronically aided bombing began. Having proved the system in the raid on the Krupps plant during Christmas week, on March 5 the RAF's Oboe-guided Pathfinder force led the first of a relentless series of raids on the factories in the Ruhr valley that would continue throughout the spring and summer. This first raid targeted the Krupps plant at Essen, and forty-two other raids followed. During that bloody campaign the RAF rained almost 60,000 tons of bombs down on the densely populated Ruhr valley, more than the Luftwaffe dropped on Britain during 1940 and 1941 combined. The tide of the bombing war was washing back into Europe, and it would soon begin to flow across Germany to Berlin.

Now it was the turn of H2S. On July 24, 1943, the RAF sent bombers to attack the shipbuilding port of Hamburg in North Germany, way beyond the reach of Oboe. It was the first major raid to be guided to its target by H2S. Hamburg was chosen because of its position at the confluence of two rivers, the Elbe and the Alster. Their smooth water would show up as a distinctive dark shape on the cathode-ray displays of the still-primitive H2S. That their town was marked out in this way was an unhappy misfortune for the people of Hamburg as almost eight hundred RAF bombers poured in across the city to bomb the flares laid by the H2S Pathfinders in what was named Operation Gomorrah. When daylight came there was no respite. This time it was the American bombers, and for two days they pounded the city. On the night of July 27, the RAF returned, once again eight hundred strong. On the twenty-ninth they came again. And four days later, on August 2. After a week of pitiless aerial bombardment three-quarters of the city was destroyed and 42,000 civilians had been killed, if not by the bombs then by the firestorm they created. Now that H2S had proved its worth, cities all across Germany became targets. Hamburg would not be the last to suffer such destruction.

By the time of the Hamburg raid the development of Mickey, the next-generation technology, was well under way back in the United States. In mid-June Rad Lab had been asked to get its own crash program going. With the assistance of the Army's own specialists and those of the Navy, Rad Lab undertook to produce twenty sets for the Army Air Forces as fast as possible.[14] The aim was to equip a single squadron of twelve planes, and to provide the necessary spares to keep the systems running in combat. The effect, though, would be hugely magnified. Just as the RAF had developed their Pathfinder concept, so now the Eighth reasoned that formation flying would allow this single squadron to lead sixty times their number on any raid. But even getting that one squadron equipped was a huge undertaking. Not only would the sets have to be designed, tested, built, installed, and tested again, there also had to be a major program of flight training, maintenance training, and another round of testing. Then the whole lot had to be flown over to England, followed by what was called "theater indoctrination." All this they achieved in a little over four months. And when they arrived the technicians were keen to do it all over again in European conditions, this time testing the systems by making simulated bombing runs over England and photographing the results. By now, though, it was already October, and with winter closing fast Mickey would make the difference between aircraft that could fly or remain grounded. Any more training they did from this point would be carried out over Germany, with sixty fully laden bombers following in their propwash.

The first such "practice" run took off on November 3, bound for the port of Wilhelmshaven. Nine Mickey-equipped planes each led sixty bombers—five hundred plus planes targeted on the docks of the North Sea port that had built, among others, the famous pocket battleship *Admiral Graf Spee*. Eight previous raids on Wilhelmshaven's docks had failed to find the target. This night, despite heavy overcast, Mickey led the bombers straight to their target. Reconnaissance photographs later showed considerable damage and concentration of bombs around the target point.

The preceding December, the Eighth had been grounded throughout the month. Now, twelve months on, Mickey facilitated the dropping of 24,000 tons of high explosive despite the fact that visual bombing was only possible on one day in fifteen. Germany's cities could no longer look forward to the safety previously afforded by overcast conditions. The era of blind bombing, "bombing through overcast," had arrived, and the nature of the air war had changed irreversibly.

Within a few weeks microwave technology would be changing the nature of war on land too. The landings in Italy agreed on at the Casablanca conference in January 1943 were launched in July with the successful invasion of Sicily. Ironically this operation was also seen as a success by the Germans, who managed to evacuate most of their troops to the Italian mainland. Sicily was followed in September by Allied landings at the foot of Italy. Despite the surrender of the Italian government that had taken over after the overthrow of Mussolini in July, the Germans decided to fight on alone. The Allies fully expected German forces now to be withdrawn north toward the Alps, but Kesselring, the German commander, persuaded Hitler that the line of defense should be drawn across southern Italy, taking advantage of the mountainous terrain. Kesselring's tactics of controlled withdrawal allowed him to build a defensive line that stopped the Allied advance in its tracks. The Winter Line ran across the country south of Rome from the Adriatic to the Tyrrhenian Sea. It was a formidable defensive position. Manned by some fifteen divisions, it was fortified with artillery emplacements, concrete bunkers, machine-gun turrets, barbed wire, and minefields. For several weeks, despite heavy fighting, particularly around the medieval monastery of Monte Cassino, the Winter Line held, until late January, when the Allies opened up another beachhead at Anzio forty miles south of Rome and behind the Winter Line.

Surrounded by a semicircle of high mountains, the flatlands at Anzio had been reclaimed from mosquito-ridden marshland by Mussolini, who had brought poor workers down from the north to help drain the marshes in return for a small plot on the new farmland. At first the Allied landings went well. They had achieved the necessary surprise and the landings were pretty well unopposed. The object of the landing had been to attack the Winter Line from the rear to force a German retreat to a point north of Rome. But instead of striking immediately, General Lucas, in command of the Allied forces, decided to consolidate his position on the beachhead. Kesselring reacted quickly, pouring troops into the surrounding mountains overlooking the Allied position and bogging down the Allied troops. By this time the invading forces were well dug in, numerous, and well equipped. But they were effectively trapped. It was a stalemate in which the two sides would pound each other in bloody artillery duels for weeks, while the Allied troops made a sitting target for Luftwaffe attacks, which now began to come thick and fast.

Into this awful deadlock came the salvation of microwave anti-aircraft gunnery. On February 24, two SCR-584s and an SCR-545 (a precursor to the superior 584 unit) were landed from Naples. By now there was a hundred square miles of beachhead to protect, and the standard long-wave radar gunnery guidance system, the SCR-268, was being effectively jammed by the Luftwaffe, who were now operating with virtual impunity. The impact of the new systems, unfamiliar to the Germans and operating on a totally unknown wavelength, was immediate and decisive. As James Phinney Baxter would record:

> On the second night after they were installed, twelve enemy bombers came over in formation, believing the eyes of

> our flak gunners had been blinded. In a few minutes half the formation came plummeting down in flames. That was the end of formation bombing at Anzio. Thereafter, enemy planes scattered before getting into flak range, resorted to violent evasive action, and bombed less accurately than before. In one night the new radars had virtually transformed defense against air attack on the beachhead.[15]

The SCR-584 had already seen action in North Africa; but it was the particular circumstances of the contained and vulnerable beachhead at Anzio that first showed its capabilities to the full. With D-Day and Operation Overlord already in the planning, Anzio was a perfect demonstration of the part that could be played by microwave-directed anti-aircraft fire in supporting newly landed troops as they tried to consolidate their position on enemy territory. The highly developed combination of SCR-584 radar, the M-9 predictor, and the newest power-driven and automatically controlled guns had shown its worth in the most dramatic style.

But there was one more job in store for the SCR-584. One week after the successful Allied landings in Normandy the Germans launched the first of their *Vergeltungswaffen*, "retaliation weapons." In Britain and America it was better known as the V-1 flying bomb, but in London, their principal target, they were called doodlebugs. For some months intelligence had been delivering a trickle of information about a German program to attack London with pilotless aircraft, and raids on prospective launch sites and factories had held the program in check for a while. The D-Day landings of June 6, 1944, though, spurred the Germans into retaliatory action, and the inhabitants of towns and villages across southeast England began to hear the distinctive drone of the primitive pulse jet that powered this prototype cruise missile. Soon they got used to it, because at first there was little to stop them coming. Gyroscopically guided, their motors were programmed to cut out when they arrived over London, creating the perverse situation that those on the ground were safe so long as the engine noise kept going. Only when it spluttered and died was it wise to throw yourself under the nearest solid shelter. Overcast or foggy conditions, which would keep regular aircraft grounded, were nothing to the V-1, and much of the time nothing could be seen as the V-1 droned overhead in thick cloud. Neither could their pilots be demoralized and discouraged by pouring flak into the airspace around them—there was no pilot to begin contemplating his own demise. If anti-aircraft guns were to prove any use against the V-1 bombs they would have to score direct hits.

This, then, was the perfect job for the SCR-584. Capable of looking through the cloud to get an exact fix on a flying bomb, they could direct an anti-aircraft gun firing shells armed with the recently developed proximity fuse with precision accuracy to explode at just the right place many thousands of feet above England shattering the incoming missile. In fact in some respects they were aided in this task by the very thing that made the V-1 so dangerous. With no pilot there could be no evasive action. The V-1 always flew straight and level, the ideal target for the anti-aircraft gunner. SCR-584s were first deployed against the V-1s in July 1944. That week they brought down 17 percent of the new flying bombs. It was just the beginning. The success rate of the SCR-584 eventually reached almost 75 percent.

Anti-aircraft guns were by no means the only defense against the flying bombs, though perhaps they were the most effective. Barrage balloons sprouted all over southern England and brought down many V-1s that got snagged by their cables. No microwaves were involved in this low-tech technique; but the cavity magnetron played a big part in another major line of defense against the V-1s. A V-1 weighed less than 5,000 pounds and could reach speeds of around 400 mph at 2,000–3,000 feet. As a target for air interception again their straight and level flight made the job comparatively simple—except that they were faster than the fighters sent up to shoot them down. Microwave radar's role in overcoming this deficiency was crucial. If they could be detected early enough on their flight path it was possible to predict their track with great accuracy and with enough time to get a patroling fighter to a point high above that track, so that when the V-1 arrived the fighter could gain speed by swooping down on to its tail. At night the flame from the V-1 exhaust would give the pilots a visual fix, but daytime intercepts were greatly assisted by onboard AI sets that could pick up the target and bring it into the fighter's cannon sights. (In some cases, an intercepting aircraft would come alongside a V-1 and use its wing to "tip" the bomb which would cause it to spin out of control and crash away from its target.) With the new generation of high-speed fighters like Hawker Tempests, Mosquitoes, the latest Spitfires, and even Gloster Meteors, the first British jet fighters in operation, these techniques produced many kills against this formidable and fast weapon.

Advanced though it was, the V-1 was something of a blunt instrument, and while the Allies were quickly able to improvise fresh defensive techniques, the attacks carried on exactly as before. The familiar chess game of ingenious defenses stimulating the introduction of counter tactics by the other side simply did not apply. The *Vergeltungswaffen* just kept on coming, come rain or shine, with London enduring four months of bombardment by V-1s before the invasion forces finally overran the launch sites in northern France and the Netherlands. While the attacks lasted, radar-controlled defensive measures were honed and improved. One warm summer Sunday at the end of August saw 105 V-1s cross the British coastline bound for London, of which only three managed to make it through to fall on a city where morale was now stronger than ever.

There was one other role that microwave radar fulfilled during this four-month rain of V-1s. Throughout 1943, Luis Alvarez and Morton Kammer, back at Rad Lab, had been perfecting what James Phinney Baxter called the "Big Bertha" of radar, the microwave early warning (MEW) system.[16] The first operational example had been shipped to Britain in January 1944, for use in the run up to the invasion of France. The MEW was a long-distance microwave search radar, with two antennae using huge reflectors to concentrate the echoes down to just 0.8 degrees. Each reflector was twenty-five feet wide. The low-coverage reflector was almost eight feet high, the high-coverage reflector just under five feet. They rotated through 360 degrees and could identify individual targets as much as 130 miles away, giving defenders the early warning the system boasted in its name. There was, however, nothing micro about the microwave early warning system. A complete MEW system weighed more than 65 tons and had to be moved in eight trucks. It would take three days, and a hundred and fifty men, to pick up and move a MEW. The electronics included five PPI displays, allowing operators to track large numbers of targets. This ability to look deep into enemy territory enabled the plotting of very precise flight paths for incoming V-1s.

One serious limitation of the MEW was that it was much better at identifying a target's bearing and distance than its height. The V-1, though, might have been designed with the MEW in mind, because its low, fixed altitude made it ideal fodder for the MEW. Once the path of an incoming V-1 was identified it could be tracked back to its point of origin, which could then be targeted by bombers with the aim of knocking out the launch equipment. Against the flying bombs the MEW proved itself to be an invaluable radar aid and, cumbersome as it was, once a second MEW had arrived in Britain by the summer of 1944 it was carried across to Normandy for use in plotting the movement of airplanes above and around the invasion armies.

D-Day brought together in a single operation all that the technological wizards had been able to devise. By this time radar had emerged as the key technology of the war for both sides. German radar had proved to be much more efficient than had earlier been thought, and the Atlantic coast, now protected by Hitler's powerful Atlantic Wall of coastal fortifications, was festooned with radar stations. Painstaking work by Britain's leading expert in electronic countermeasures, R. V. Jones, had identified virtually every single German radar installation along the Atlantic coast—some two hundred of them. Three weeks before D-Day, on May 10, the fourth anniversary of Hitler's invasion of the Low Countries, Allied bombers began a concerted campaign to knock them out ahead of the invasion. Deception and secrecy were central to the success of the operation, and so for every one station attacked in the Normandy landing area, another two were attacked on other sectors of the Atlantic coast. A single site, at Fècamp, twenty-five miles north of Le Havre in Brittany, was deliberately left untouched. There had to be at least one site in operation on the day to pick up the false information the Allies intended to sow.

The task of knocking out targets to reduce the enemy's ability to strike back at invasion forces was facilitated by a period of exceptionally fine weather, allowing visual aiming by the Eighth Air Force; but at night Oboe and GEE were used to pinpoint bridges, rail marshaling yards, and coastal batteries. On the night before the invasion, the fine weather disappeared and visual bombing became impossible. Protecting the beaches where thousands of men would come wading ashore was a string of eleven German coastal batteries. Taking them out too early would have signaled the location of the intended invasion. Now they were themselves protected by a cloak of heavy skies—accurate blind bombing would be needed to destroy them during the dark night, and just hours before the landings. MkIII Oboe, the microwave version of the system that had devastated the Ruhr, was used to guide the heavy bombers that night, and not one of the batteries fired a single shot during the landings the following day.

At Dunkirk, four years earlier almost to the day, hundreds of thousands of Allied troops had found themselves marooned on broad Atlantic beaches, easy prey for the Luftwaffe planes speeding along the dunes. Now, they were once again to find themselves anxiously scanning the skies for German planes as they assembled on the sands of Normandy. By June 1944, the Luftwaffe was no longer the force it had once been, but it was still capable of inflicting terrible damage on such a target. Anti-aircraft guns were therefore among the first equipment to be unloaded from the landing craft that morning. Altogether no fewer than thirty-nine SCR-584s went into action on the shores of France that day protecting the multinational force from the attention of the Luftwaffe. A little less than four years had passed since the first cavity magnetron crossed the Atlantic in a plea for help. Now it was crossing the English Channel to Europe in its many forms ready for the drive on to Berlin.

EIGHTEEN

The Mission's Legacy

IN THE FALL OF 1940, AS BRITAIN STOOD ALONE and faced imminent invasion, Tizard's mission was all about the development and production of weapons required at that moment. But unusually the British were also thinking about weapons that were still only a theoretical possibility. Well before Tizard climbed on board the flying boat *Clare* for his historic trip, he was involved in the early developments in atomic physics that would open the way to the A-bomb, and the history of atomic weapons followed a path strangely parallel to that of microwave radar.

Late in 1938 the German Nobel Prize-winning chemist Otto Hahn and his assistant Fritz Strassman succeeded in achieving nuclear fission for the first time. Just before Christmas they communicated their findings to *Naturwissenschaften*, the house journal of the Kaiser Wilhelm Institutes. At the same time they sent the details to Hahn's former collaborator Lise Meitner, who had escaped from Germany to Sweden earlier that year. Meitner went over them with her nephew, the young physicist Otto Frisch, and it was Frisch who coined the term "nuclear fission" for what Hahn and Strassman had been able to achieve. Frisch confirmed the results experimentally on January 13, 1939.

By the fall of 1939, the possibility of this discovery laying the groundwork for the production of a bomb had so galvanized some of the key scientific figures now fleeing the Nazi regime in Europe that Leo Szilard, a Hungarian physicist who had first conceived the possibility of a nuclear chain reaction back in 1933, drafted a letter to President Roosevelt. It was signed by, among others, Albert Einstein, and it urged that the U.S. should embark on the development of a nuclear weapon, on the grounds that Germany would almost certainly already be doing so. At this stage though, the weapon envisaged would have been huge. A ship would be needed to deliver it, yet theoretically it was capable of destroying any major port city.

It was only in the spring of 1940 that this limitation was shown to be spurious. By this time Frisch had arrived in Britain as a refugee from an enemy state; he was not allowed to involve himself in the vital war work of radar, but instead was given the facilities, along with a fellow refugee from Germany, Rudolf Peierls, to carry on more theoretical studies. In March 1940, they published a memorandum that changed everything. Just three pages long, its key conclusion was that: "a moderate amount of U-235 would indeed constitute an extremely efficient explosive."[1] Nuclear warfare became suddenly feasible. The two men had fundamentally revised the calculations about the size of the critical mass needed for an atomic bomb, reasoning that it would require only about one pound of the isotope uranium-235. (In the event the amount needed was somewhat greater, but the principle was sound—and revolutionary.) By extraordinary chance they had been invited to carry out this work under the auspices of Mark Oliphant at the University of Birmingham, just a few yards away from where Randall and Boot were building their first cavity magnetron at almost exactly the same time.

Oliphant was quick to see the significance of the document, and he wanted Tizard to know all about it as soon as possible. Soon Frisch and Peierls were on their way to London to discuss the matter with Tizard.[2] A new committee, codenamed MAUD, was convened by Tizard to pull together all the research into the feasibility of building an atomic bomb. For a little over a year the MAUD Committee coordinated all the work going on in the UK together with the wealth of expertise now available there through the many refugee scientists pouring in from Europe to escape the Nazi yoke. The French in particular had been advancing work on the use of nuclear fission for power generation, and key scientists, including Hans von Halban and Lew Kowarski, were spirited away to Britain as the German armies descended on Paris, together with an invaluable supply of heavy water, enriched with deuterium. In these early days of nuclear research heavy water was a vital commodity, and when the 8,000 ton collier the *Broompark* sailed from Bordeaux the most valuable part of its cargo was neither £2.5 million-worth of industrial diamonds, nor some forty of France's top scientists, but thirty-six gallons of heavy water. The heavy water was fixed to a raft so it would float in the event of the *Broompark*'s accidental sinking, while at the same time being chained to the ship so it would sink if it were torpedoed by a German U-boat. This curious cargo was well worth saving—those twelve cans of heavy water comprised most of the stock of the entire world. Had it fallen into the hands of the Germans it could have considerably accelerated Nazi nuclear research.

From the start it was decided that the research available to the British should be made available across the Atlantic, and in July 1940, the first deliberations of the MAUD committee were communicated to R. H. Fowler, Britain's scientific attaché in Canada, so that he could pass them on through his extensive and regular contacts in Washington. This one-way flow of information was kept up for a full twelve months until, in the summer of 1941, the MAUD Committee produced two substantial reports drawing all the currently available science together. The committee had reached the conclusion that this new "super weapon" could indeed be realized before the war was over, and that the entire research effort should therefore be diverted from non-military uses of the technology into developing a bomb. At the same time the committee was convinced that the vast resources needed, both human and financial, were beyond the United Kingdom at a time when she was facing the imminent prospect of defeat.

All this was in stark contrast to the leisurely pace at which things were unraveling in the United States, where a mere $6,000 had so far been allocated to nuclear research. In March 1941, the Advisory

Committee on Uranium, set up after Roosevelt received the Szilard letter, learned a key fact from the British. New calculations recently carried out by Rudolf Peierls showed the correct critical mass (the smallest amount of material needed for a sustained nuclear chain reaction) for U-235 to be 18 pounds as a bare sphere, or 9 to 10 pounds if a reflector were used. It was vital information, but seemed to provoke virtually no response from the American scientists. When John Cockcroft returned to Britain from heading the Tizard mission after Sir Henry's departure he reported that the work going on in America was "very much along the same lines as our own, but perhaps not so well advanced."[3] It was diplomatic language that concealed a desultory approach by the Americans in sharp contrast to Britain's keen awareness that, with the right resources applied to the job, a new weapon could be available in time to affect the outcome of the war.

The U.S. did not, of course, yet have the adrenaline of war pumping through its veins. But by late summer 1941, its entry into the conflict was beginning to look inevitable. Toward the end of August Marcus Oliphant arrived in the United States, ostensibly to check progress on the radar program, but with a secret agenda to push the Americans on the manufacture of an atomic weapon in the wake of the MAUD reports, and to find out why they seemed to be ignoring the research they had been passed by the committee, of which he was a member. Oliphant reported back that, "The minutes and reports had been sent to Lyman Briggs, who was the Director of the Uranium Committee, and we were puzzled to receive virtually no comment. I called on Briggs in Washington, only to find out that this inarticulate and unimpressive man had put the reports in his safe and had not shown them to members of his committee. I was amazed and distressed."[4]

Disappointed in the director, Oliphant arranged to meet with Samuel K. Allison, a new member on the Uranium Committee. His aim was to galvanize the committee from the floor up. According to Allison, Oliphant said that the Uranium Committee "must concentrate every effort on the bomb and. . . had no right to work on power plants or anything but the bomb. The bomb would cost 25 million dollars. . . and Britain did not have the money or the manpower, so it was up to us."[5] Oliphant carried on doing the rounds of major scientific figures in the U.S. to explain the urgency of developing the bomb before the Nazis could. He visited his old friend Ernest Lawrence, along with James Conant and Enrico Fermi (who would go on to create the world's first functioning nuclear reactor in December 1942). Even Vannevar Bush had been lukewarm on the prospects of developing a bomb, but when Roosevelt transformed the NDRC into the more powerful (and congressionally funded) Office of Scientific Research and Development (OSRD) he was empowered to engage in large engineering projects in addition to research. This reorganization, together with the work handed over by the MAUD Committee and the energizing effect of Oliphant's visit, opened the way to the wholesale application of American resources, know-how, and productive capacity to a major military objective. The memorandum by Frisch and Peierls and the MAUD Committee's reports were both major milestones in stimulating America to take up the race to produce an atomic weapon. Without them it is doubtful that America would have had atomic weapons with which to try to accelerate the end of the war in Japan. Sums of money that would have seemed unthinkable just a year before were now available to research and development teams, including an "all-out" program with the Uranium Committee.[6]

On December 6, 1941, with America now just a day away from being drawn into the war, the uranium section was not only given the green light but told they were henceforth to come under the control of James B. Conant as the representative of Vannevar Bush himself—Briggs had been relieved of responsibility. The Radiation Laboratory, still less than a year old but already changing the face of the radar war, was about to be joined by a parallel project of equally staggering proportions and impact—the Manhattan Project.

It is a strange curiosity that of the two major breakthroughs at Birmingham University in the early months of 1940, the cavity magnetron should have so dominated the Tizard mission, while the new feasibility of an atomic bomb scarcely figured. All the more so as Cockcroft, Fowler, and Hill were among the world's foremost physicists and, along with Tizard, were fully aware of all the advances in nuclear theory. The explanation may lie in a combination of cautious skepticism and the short-term imperatives of war. In the spring of 1940 the Frisch and Peierls memorandum, brilliant as it was, awaited independent substantiation. There seems to have been a reluctance, even at the highest scientific levels, to contemplate the possibility of such awesome power coming into the hands of human beings. The mission itself had only three meetings on nuclear matters. The most significant was on October 7, when Cockcroft and Fowler met Lyman Briggs's committee. Cockcroft's report of the meeting records that "in general the Committee considers this to be a long term programme."[7] It was a view evidently shared by the British—and for Britain in the fall of 1940 there simply was no "long term."

Nevertheless, once the possibilities had finally been recognized, and the fission project had been taken over by Bush and the OSRD, it was clear that there was no need to look any further for the Manhattan Project's core staffing than Rad Lab. On both sides of the Atlantic, radar had sucked in most of the brightest physicists. By mid-1943 the development phase of microwave work at Rad Lab had been overtaken by the urgency of production. Some 6,000 sets had already been supplied to the military, and there were a further 22,000 on order. Four thousand people now worked at Rad Lab, and five hundred of them were physicists—many of them keen to take on the next major challenge.

J. Robert Oppenheimer was not one of those men. He joined the Manhattan Project independently as a theoretical physicist, working on fast neutron calculations. He had studied at both the Cavendish Laboratory in Cambridge, England, where he was tutored by Patrick Blackett, and then at the University of Göttingen under Max Born. Göttingen was one of the world's leading centers for theoretical physics, and here he met Werner Heisenberg, Enrico Fermi, and Edward Teller, among many others. Theoretical physics was a small world in the 1920s and 1930s. Returning to the U.S. he had taken up a position first at Caltech and then Berkeley, working alongside Ernest Lawrence. By the time war exploded in Europe Oppenheimer was already deeply immersed in work on nuclear fission. By 1942, when he was appointed scientific director of the Manhattan Project, his extensive network of contacts, and the high intellectual regard in which he was held, opened the door to a rapid flow of talent from Rad Lab to a new laboratory at Los Alamos, New Mexico.

Understandably, DuBridge and Loomis were none too pleased that their brightest minds were being enticed away; but they soon bowed to the inevitable and watched a long list of Rad Lab high achievers leave for the top-secret project in New Mexico. George Kistiakowsky, Norman Ramsey, Luis Alvarez, Bob Bacher, Hans Berthe—all these and more were Rad Lab's gift to the team who would develop the atomic bomb over the next couple of years. Each of them was aware of the terrible responsibility they were taking on, but for the vast majority the battle against fascism and totalitarianism fully legitimated what they were about to do. A few, including the highly principled Rabi, could not bring themselves to join the Los Alamos team, staying at Rad Lab and acting only as a consultant to the atomic work, almost uniquely moving freely between the two.

The worlds of radar and nuclear physics overlapped in many ways during the war and the few years leading up to it. Both were huge undertakings that pushed the boundaries of engineering, and both were built on the genius of a handful of physicists, many of whom crossed from one project into the other. Many of them, too, had passed through Tower House, Loomis's private laboratory at Tuxedo Park. It was a place where the prewar internationalism of the scientific community was fostered and developed and where ideas could be shared and extrapolated. And it was the birthplace of Rad Lab. But as Rad Lab's work matured into the production phase, and the electronics companies began turning out microwave radar equipment in huge quantities, while many of the key scientists involved moved on to the Manhattan Project, Alfred Lee Loomis himself seems to have disappeared from the story.

On April 4, 1945, Alfred Lee Loomis and his wife Ellen were divorced. A few hours later, in Carson City, Nevada, he married again, to a much younger woman and the former wife of a colleague. Loomis was generous in the financial settlement with his wife, making over more than half of his considerable fortune along with the house in Tuxedo Park and their New York penthouse. He bought a large serviced apartment in Manhattan and a summer cottage in the Hamptons. It was here that the newlywed couple discovered the full extent of the ostracism that awaited them in days when divorce was a scandal all by itself. The society that had once formed the bedrock of Loomis's charmed existence now turned its back on him, and the couple melted away into mutually supportive obscurity. Rad Lab would suffer a similar fate.

It is often said that wars are not won by committees. Sir Henry Tizard was, however, the ultimate committee man, capable of managing the complex dynamics inherent in committee work as well as inspiring great loyalty on the part of his fellow members. The Tizard mission was in many senses a committee created by a committee, and it bore all the hallmarks of Tizard's particular skills. Nevertheless, it is true that most wars are driven by individuals who take to themselves the immense powers of the state. Such a person requires great charisma and single-mindedness, to the point of stubborn obsession. Such a man was Winston Churchill, and at his side, from May 1940 when he arrived at 10 Downing Street onward, was another—Frederick Lindemann, known one suspects with little affection as "the Prof." After Tizard set the mission on its way in the fall of 1940 he came back to a world where all the key scientific decisions were being filtered through Lindemann, his principal detractor. And with Lindemann where he was, there was little chance of Tizard, perhaps Britain's most skilled military scientific administrator, being offered a position of power. Soon after his return Tizard became a semi-official adviser to the Minister of Aircraft Production, sitting on the Aircraft Supply Committee and representing the ministry on the Air Council from 1941. Here he was particularly helpful in maintaining the exchange of information with Washington. In 1942 he was made chairman of a committee dealing with radar and radio jamming techniques, an area of growing significance in the increasingly technological war. But once again the old animosities between Tizard and Lindemann surfaced, and Tizard was forced to step down from the chairmanship.

This spat was dwarfed, though, by a major row over strategic bombing that broke out in 1942 following the publication of the Butt Report. Lindemann, with the support of Arthur Harris, newly installed as head of the RAF's Bomber Command, proposed a change of tactics aimed at demoralizing the German public by quite simply bombing them out of their homes. It was a policy enshrined in the Area Bombing Directive of February 14, 1942, issued to Bomber Command, stating, "the primary objective of your operations should be focused on the morale of the enemy civil population and in particular the industrial workers."[8]

Lindemann, who prided himself on his statistical prowess, had produced in evidence a paper estimating what could be achieved by area bombing. Tizard, with the support of Patrick Blackett, chief scientist to the Royal Navy, disagreed, saying Lindemann's estimate of being able to achieve 50 percent destruction of targeted German cities was way too high—by as much as tenfold. On April 15, Tizard wrote to Lindemann questioning the mathematics behind the paper and warning that the War Cabinet could reach the wrong decision if they based their decision on his paper. Tizard maintained that the only benefit of strategic area bombing was that it tied up enemy resources defending Germany. He believed that those forces could be occupied just as well with a far smaller, more focused bombing offensive, thus releasing RAF resources that could be put to better use elsewhere. It was a debate that went right to the heart of policy for the rest of the war. Tizard lost, and so did the people of Hamburg, Cologne, and Dresden.

In the aftermath of this bruising row Tizard was offered the post of president of Magdalen College, Oxford. Though ostensibly merely swapping one academic post (his rectorship at Imperial College) for another, in practice he was stepping away from involvement in the management of the war. He continued to watch from the sidelines until the end of the war, when the British public unceremoniously threw Churchill out of office in favor of the new order promised by Labour, Britain's socialist party, and the incoming government. Peace is no respecter of wartime reputations.

James Phinney Baxter won the 1947 Pulitzer Prize for *Scientists Against Time*, his official history of the Office of Scientific Research and Development. In a comprehensive analysis of all the wartime achievements of this substantial organization he described the cavity magnetron as "the most valuable cargo ever brought to our shores."[9] Radar and microwaves made Rad Lab into a large and astoundingly productive business. By the end of the war the United States had spent a total of more than $2.8 billion on radar, very nearly 50 percent of it designed by Rad Lab. It changed the face of World War II. But what was its legacy?

In some ways it has been usurped in military usage by newer developments—satellite technologies and global positioning. Yet cavity magnetrons are still in daily use all across the world because the microwave oven is powered by a cavity magnetron. The technology that burst into life on a laboratory bench in Birmingham, England in February 1940 and went on to power the most sophisticated military systems in the great struggle against fascism now does service warming up TV dinners and making popcorn.

At the time of this writing there are an estimated 600 million microwave ovens in use across the globe—600 million cavity magnetrons powering them with exemplary reliability and safety. The design, of course, has moved on, but the principle of operation is exactly the same as that built by Randall and Boot more than seventy years ago. Of how many inventions can that be said? Though perhaps less visible, the cavity magnetron is worthy of consideration alongside a few (*the* few) other inventions that have stood the test of time in the same way. The incandescent light bulb, television, radio, the internal combustion engine—each has been refined over the years, but few have retained their fundamental form to the same degree as the magnetron.

The magnetron still has some military applications, in the field of missiles and flight control, but in the civilian world it offers low cost, high power, and reliability: a magic combination. Today it remains the prime radiowave power source of civil marine radar and coastal control. It is used to control ground movements at airports and for weather radar, for cargo screening and for plasma generation in the making of artificial diamonds. It is a vital component of industrial cooking and a wide variety of other industrial processes. It is even being used in cancer treatments.

And as its power and reliability grew, so the cost has dropped. What was once transported in utmost secrecy across the Atlantic in Tizard's deed box, and cost the princely sum of $1,000 when it appeared in the first microwaves ovens in the 1950s and 1960s, is now a mere $10. It is ubiquitous and resilient—and immensely powerful. It can happily work with pulses of up to four kilovolts (one kilovolt, a thousand volts, is enough to power an electric light rail system). Yet we still don't quite understand what is happening inside it. All the hundreds of millions of magnetrons all over the world function in fundamentally the same way as that original small, circular lump of metal with tubes protruding at either end. It was a wartime secret so special that even the stenographer who accompanied the mission to the United States did not know about it at the time: "Not until several years after the mission ended did I know that the leather suitcase I was forced to keep under my bed contained the prototype of the cavity magnetron."[10]

Yet how necessary was all that secrecy? In war, as in comedy, timing is everything—along with occasionally making the correct guess. The Dutch historian Arthur O. Bauer has made a study of German radar, and in particular what happened in Germany in relation to microwaves and centimetric radar. It is clear that German scientists were aware of Russian work on a cavity magnetron and that the production of microwaves at high power was more than a theoretical possibility. In Russia, Alekseev and Malyarov, working under their director Bonch-Bruevich, had published a paper in April 1940 detailing the work they had done to build a four-cavity magnetron. But though they had built one in a lab, it had never been put into production. Bonch-Bruevich had died a month before the paper was published, and Malyarov would be killed in the siege of Leningrad in 1942. Czech scientists, too, had been on the way to unlocking the secret of the magnetron, until Hitler closed all Czechoslovakia's universities in September 1939. Even the Japanese had a prototype microwave radar system operating as early as 1928—though they did not have the cavity magnetron to power it. The Germans already had *Knickebein* and some highly developed radar systems operating at meter wavelengths. So they would have been well aware of the benefits of microwave radar. But the question was, how long might it take to get it working and available in the front line? Berlin, of course, never planned for a long war. Nazi tactics were about finishing things off quickly, and that's exactly how it had worked out until the Russian campaign ground to a halt outside Moscow in the winter of 1941–42. Now, suddenly, things weren't going so well and a sense of desperation began to creep in alongside the desire to find an immediate solution. Bauer points out that, just as Eddie Bowen had been told in Britain, microwaves were not for this war:

> As the burden of war became ever heavier Germany started rationalizing. A "Führer Befehl" was issued, which decreed that every project, not guaranteed to be ready for the front within six months, had to be abolished. Dr. Runge, head of radar development at Telefunken, saw no use for centimetre waves in radar, as . . . the amount of energy coming back in the direction of the radar set was considered to be too small to be useful. So the decision to terminate development of centimetre radar was easily taken.[11]

Less than three weeks later Telefunken engineers recovered the microwave-powered H2S system from the Stirling shot down near Rotterdam. Microwaves were clearly for this war after all.

What Britain had, that others didn't, was the *need* for microwave radar. Because it was fighting a defensive war in 1940, and an aerial one at that, and because Tizard had the foresight to anticipate the need to deal with bombers coming singly at night, airborne interception was named the number one technological priority of the war. Randall and Boot, motivated by their country's circumstances and brought together by inspired leaders, came up with the means to make it possible. The Tizard mission put it in the hands of those who could provide the necessary resources, and the microwave future was under way. But it could all have been so different.

According to Lovell, a German correspondent who was present at the dinner at the observatory in Effelsberg in 1977 recalled that Hachenberg said he had been told that the command to drop microwave research and concentrate on existing meter-wavelength technology had been by “higher order.” (The correspondent, Professor Wielebinski, speculated that this might have meant Hitler himself.)[12] At a very early stage of development, Lovell recalls, just such an order came down from on high in the UK. A similar order was given to A. P. Rowe (the superintendent of the TRE) as the German armies surged across France in the spring of 1940. At that moment it seemed most improbable that any research or development of centimetric radar could possibly have any relevance to a fast-paced war that threatened to reach its climax in weeks or months rather than years, and that all possible effort should be concentrated on the installation, maintenance, and use of the meter-wave radar already available.

At that critical moment W. B. Lewis, Rowe’s deputy, persuaded him not to enforce the order on the group of half a dozen newly recruited scientists working on centimetric developments in a small wooden hut remote from the main establishment at Worth Matravers.[13] Of such small decisions is history made.

Quite how much of a huge leap the cavity magnetron was is difficult to convey, spawning as it did so many new systems in so many different fields. Its mark on our lives today is massive, but less visible than the fruits of the computer revolution, for instance. Perhaps it is best to leave the task to the people who lived through it all, who made it happen. More than forty years after the Tizard mission, Luis Alvarez had this to say about the difference the cavity magnetron made: “A sudden improvement by a factor of three thousand may not surprise physicists, but it is almost unheard of in engineering. . . . If automobiles had been similarly improved, modern cars would cost about a dollar and go for a thousand miles on a gallon of gas.”[14]

And A. P. Rowe, who was there at the very beginning, writing shortly after the war had ended, was clear that the cavity magnetron, and the people who worked out how to use it, may well have ensured the survival of the liberty that British and Americans alike so cherished: “We used to say, with some truth as well as egotism, that if TRE were in Germany the work of Bomber Command would be impossible. What is more nearly true is that if the Service user and the radar scientist had got together in Germany as they did in this country, the war might well have taken a different course.”[15]

In America they *did* get together in that way, and to huge effect. That they did so is largely due to the influence of Tizard and his mission. It is fitting to leave the last word to the man himself, Sir Henry Tizard, writing in 1946: "When I went to Washington in 1940 I found that radar had been invented in America at about the same time it had been invented in England. We were, however, a very long way ahead in its practical application to war. The reason for this was that scientists and serving officers had combined before the War to study its tactical uses. This is the great lesson of the last war."[16]

3. The Chairman emphasised that it was quite likely that we should give more information than we received, but the object of the mission was largely to create goodwill. In general, he proposed to pass on full information about matters which had already reached a fairly advanced stage of development, but only very general indications of the lines on which we were working on subjects which were still in the very early stages.

4. D.S.R. Admiralty said that the view of the Controller was that we should be forthcoming on all subjects on which information was definitely asked, but that that we should not volunteer information on very new developments. Sir Henry Tizard said that the Controller had verbally agreed that he should be empowered to call upon any of the Admiralty representatives in the U.S.A. or Canada, but that the official Admiralty letter mentioned only the name of Commander Crossman. D.S.R Admiralty agreed to take this matter up with the Controller.

5. Sir Henry Tizard said that he had obtained a list of all service representatives in the USA and Canada, and he was obtaining notes of subjects on which each was specially qualified to help.

6. The Meeting now considered specific matters as tabulated below:

GUN TURRETS IN AIRCRAFT

It was agreed that full descriptions and drawings of our turrets should be taken, which would be handed over to the authorities if wanted. Air Commodore Baker (lately D.Arm.D.) would be available in the U.S.A. as a specialist on the subject.

R.D.F

An A.S.V. and I.F.F. equipment would be sent to Canada with a view to fitting in a Hudson and demonstrating its working, in the U.S.A. Professor Cockcroft was also arranging for other R.D.F. equipment to be sent over.

ROCKET DEFENCE OF SHIPS

It was agreed to inform the Americans of this emphasising that it was being used largely owing to an insufficient supply of guns. Professor Cockcroft agreed to produce photographs of the apparatus and a slow motion film, subject to the agreement of the Controller.

MULTIPLE POM-POM

Captain Faulkner would be able to deal with this. Photographs would be supplied if possible.

ARMOUR PLATE

Professor Cockcroft agreed to obtain information on the latest position and ascertain the names of experts already in the USA. Full information would be given, including that on the armouring of the decks of aircraft carriers.

LIGHT A.A. GUNS

Professor Cockcroft would provide details of the Kerrison Predictor and other equipment, also criticisms of the Sperry method. We should ask the Americans for details of the Sperry type bombing predictor.

CHEMICAL WARFARE

That could better to be dealt with Colonel Barley was already in the USA.

EXPLOSIVES

The Americans probably already had details of R.D.X. They might be able to help in getting it made. It was suggested that Captain Priston (lately C.S.R.D) might be invited to accompany the Mission as a further Admiralty representative.

GYRO PREDICTOR GUN-SIGHT

It was agreed that the problem should be mentioned and an indication given of the general lines on which we were tackling it. If the Americans wanted to send a representative over to this country to see what we were doing, this should be encouraged.

PETROL INJECTION ENGINES

Dr. Ricardo has supplied information to the Chairman.

JET PROPULSION ENGINES & INTERNAL COMBUSTION TURBINES

D.S.R. (M.A.P.) has supplied details of the position in general terms. No technical details will be given.

AUTOMATIC OXYGEN SEPARATOR

A general indication only would be given, since this was in the very early stages.

MARTIN BAKER BARRAGE CABLE CUTTERS

A sample with descriptive matter, and, if possible, a film would be taken.

D.P.L. SCHEME

Samples will be taken.

P.A.C. SCHEME

Units and if possible a film, with descriptive matter, would be taken.

LONG AERIAL MINES

A sample and if possible films will be taken.

MAGNETIC MINES (NAVAL)

D.S.R. Admiralty was not anxious that information on the latest pattern should be disclosed, if not already in the possession of the Americans. It was pointed out, however, that it had already been disclosed to the French and that information was therefore probably in the possession of the Germans.

GLIDING TORPEDOES

A statement of the present position in general terms should be provided, if possible accompanied by photographs or films.

RADIO NAVIGATION

Dr. Bowen would be able to speak on this, and details of our work and also of the German system would be given.

COMPASS

A specimen of a captured German distant reading compass should be taken if possible, together with descriptive matter.

SELF SEALING PETROL TANKS

Descriptions of both the German and our tanks should be given and samples taken, if possible.

1/2″ GUN FOR AIRCRAFT

Group Captain Pearce was to discuss the operational aspects and ascertain whether there was a requirement for this type of gun.

7. In general, it was agreed that as many films and as much descriptive matter of various new devices should be taken. Professor Cockcroft agreed that the Ministry of Supply (Dr. W.L. Webster) would undertake the responsibility of seeing that all samples, films, etc got over, if they were sent to him in the first instance. It was agreed to write to the Ministry of Supply on these lines.

8. Arrangements for passages etc. would be made by the M.A.P.

Notes

Chapter 1

1. Ronald W. Clark, *Tizard* (London: Methuen, 1965), 13.

2. The case for Germany's scientific pre-eminence in this period is eloquently made by John Cornwell, in his clinical analysis of where it all went wrong in *Hitler's Scientists* (London: Viking, 2003), 40.

3. Sir Henry Tizard, Unfinished Autobiography, quoted in Clark, *Tizard*, 16.

4. Basil Collier, *The Battle of Britain* (London: Batsford, 1962), 18.

5. *The New Cambridge Modern History*, Vol. XII (Cambridge: Cambridge University Press, 1968), 283.

Chapter 2

1. U.S. declaration of neutrality, August 19, 1914.

2. Henry Cabot Lodge, "The League of Nations Must Be Revised," *Congressional Record*, 66th Cong., 1st sess., 1919, 3779-84.

3. John E. Wiltz, *In Search of Peace: The Senate Munitions Inquiry, 1934–1936* (Baton Rouge: Louisiana State University Press, 1963) pp. 15, 69, 73-74, cited in Geoffrey S. Smith, "Isolationism, the Devil, and the Advent of the Second World War: Variations on a Theme," International History Review 4, no. 1 (Feb. 1982): 62.

4. Franklin D. Roosevelt, *The Public Papers and Addresses of Franklin D Roosevelt*, 2nd series, ed. Samuel I. Rosenman, 13 vols. (New York: Russell and Russell, 1941-50), 8:1565.

5. *Franklin D. Roosevelt and Foreign Affairs.* 1st series, ed. Edgar B. Nixon, 3 vols. (Cambridge Mass.: Belknap Press, 1969), 3:251, quoted in Steven Casey, *Cautious Crusade* (New York: Oxford University Press, 2001), 6, 229 n.7.

6. Franklin D. Roosevelt, *FDR: His Personal Letters*, ed. Elliott Roosevelt, 2 vols. (New York: Duell, Sloan, and Pearce, 1950), 1:716-717, quoted in Casey, *Cautious Crusade*, 23.

Chapter 3

1. A. P. Rowe, *One Story of Radar* (Cambridge: Cambridge University Press, 1948), 45.

2. Quoted in Clark, *Tizard*, 111.

3. Rowe, *One Story of Radar*, 5.

4. Quoted in Clark, *Tizard*, 113.

5. Quoted in Clark, *Tizard*, 113.

6. Quoted in Clark, *Tizard*, 115.

7. Wimperis's diary, quoted in Clark, *Tizard*, 118.
8. Draft of speech, Cherwell Archives, Nuffield College Oxford F7/2/1.
9. Clark, *Tizard*, 123.
10. Draft letter to Tizard, Cherwell Archives, Nuffield College Oxford F6/5/5.
11. Draft letter to Tizard, Cherwell Archives, Nuffield College Oxford F6/5/10.

CHAPTER 4

1. Alan Bullock, *Hitler: A Study in Tyranny* (London: Odhams, 1952), 135.
2. Churchill to Swinton (June 22, 1936), Cherwell Archives, Nuffield College Oxford F8/1/12.
3. Letter to Tizard (June 25, 1936), Cherwell Archives, Nuffield College Oxford F8/1/10.
4. Tizard to Lindeman (July 5, 1936), Cherwell Archives, Nuffield College Oxford F8/6/6.
5. Lindemann to Swinton (September 23, 1936), Cherwell Archives, Nuffield College Oxford F8/5/12
6. Swinton to Lindemann (November 2, 1936), Cherwell Archives, Nuffield College Oxford F8/5/13.
7. Sir Henry Tizard, Unfinished Autobiography, quoted in Clark, *Tizard*.
8. E. G. Bowen, *Radar Days* (Bristol: Adam Hilger, 1987), 30.
9. Bowen, *Radar Days*, 45.
10. General Account of Army Radar, by Prof. J. D. Cockcroft, Tizard Papers, Imperial War Museum, London HTT 701.
11. General Account of Army Radar, by Prof. J. D. Cockcroft, Tizard Papers, Imperial War Museum, London HTT 701.
12. General Account of Army Radar, by Prof. J. D. Cockcroft, Tizard Papers, Imperial War Museum, London HTT 701.
13. Conversation with Sir Mark Oliphant, July 24, 1967, National Library Collection, Tape 276, p. 5 of 12 page transcript (Interviewed by Hazel de Berg). Published by the Australian Science Archives Project on ASAPWeb, June 26, 1996.
14. General Account of Army Radar, by Prof. J. D. Cockcroft, Tizard Papers, Imperial War Museum, London HTT 701.
15. Author interview with Dr. Phil Judkins, Defence Electronics History Society.

CHAPTER 5

1. G. C. Pirie, "Visit to Infantry School, Fort Benning, Columbia, Georgia, on April 13th to witness Air Demonstration," Public Record Office, AIR 2/3339, quoted in David Zimmerman, *Top Secret Exchange*, (Montreal: McGill-Queens University Press, 1996) , 37.

2. Roosevelt to Chamberlain, 31 August 1939, Public Record Office, AIR 2/3339.
3. Winston Churchill, *The Second World War, Vol. II, Their Finest Hour* (London: Cassell, 1948), 354.
4. Letter from A. V. Hill, Tizard Papers, Imperial War Museum, London, HTT 706.
5. Letter from A. V. Hill, Tizard Papers, Imperial War Museum, London, HTT 706.
6. Letter from A. V. Hill, Tizard Papers, Imperial War Museum, London, HTT 706.
7. Letter from A. V. Hill, Tizard Papers, Imperial War Museum, London, HTT 706.
8. Bowen, *Radar Days*, 151.
9. Letter from A. V. Hill, Tizard Papers, Imperial War Museum, London, HTT 706.
10. Telegram from Lord Lothian to Foreign Office. Public Record Office, FO 371/24255.
11. Ibid.
12. J. Balfour to F. H. Sandford, Air Ministry, 24 April 1940. Public Record Office, FO 371/24255.

Chapter 6
1. Winston Churchill, *The Second World War, Vol. I, The Gathering Storm* (London: Cassell, 1948), 667.
2. Churchill, *Their Finest Hour*, 88.
3. J. G. Crowther, *Statesmen of Science* (London: Cresset Press, 1965), 340.
4. R. V. Jones, Scientific intelligence of the Royal Air Force, quoted in *The Conduct of the Air War in the Second World War*, ed. Horst Boog (New York: Berg, 1992), 583.
5. Churchill, *Their Finest Hour*, 339–340.
6. Churchill, *Their Finest Hour*, 339–340.
7. *Royal Society: Papers of PMS Blackett*—Tizard, H.T.; Correspondence of various dates CSAC 63.1.79/J.105 1940–1958.

Chapter 7
1. Churchill, *Their Finest Hour*, 23.
2. Quoted in Joseph P. Lash, *Roosevelt and Churchill 1939–1941* (London: Andre Deutsch, 1977), 152.
3. *Public Papers and Addresses of Franklin D. Roosevelt* (New York: Macmillan, 1941), 263–264 (quoted in Lash, 152).
4. Quoted in Lash, *Roosevelt and Churchill*, 357.

5. William Langer and S. Everett Gleason, *The Challenge to Isolation* (New York: Harper, 1952), 521–522.
6. Churchill, *Their Finest Hour*, 198.
7. Churchill, *Their Finest Hour*, 201.
8. Lash, *Roosevelt and Churchill*, 161.

CHAPTER 8
1. Bowen, *Radar Days*, 142.
2. Bowen, *Radar Days*, 143.
3. Bowen, *Radar Days*, 143.
4. On this flight by zeppelin LZ 130, see Robert Buderi, *Invention that Changed the World* (New York: Simon and Schuster, 1996), 91.
5. Rowe, *One Story of Radar*, 56.

CHAPTER 9
1. James Phinney Baxter, *Scientists Against Time* (Boston: Little, Brown, 1946), 14.
2. "Order Establishing the National Defense Research Committeee," http://docs.fdrlibrary.marist.edu/psf/box2/a13v01.html.
3. The final figure was Democrats 261, Republicans 164.
4. Quoted in David Reynolds, *From Munich to Pearl Harbor* (Chicago: Ivan R. Dee, 2001), 45.
5. http://en.wikisource.org/wiki/Roosevelt%27s_Fireside_Chat,_3_September_1939.
6. http://www.firstworldwar.com/source/usneutrality.htm
7. Quoted in Reynolds, *From Munich to Pearl Harbor*, 71.
8. Lash, *Roosevelt and Churchill*, 113.
9. Casey, *Cautious Crusade*, 11.
10. Isaiah Berlin, *Mr. Churchill in 1940* (Boston: Houghton Mifflin, 1960), 18.
11. Berlin, *Mr. Churchill in 1940*, 28.

CHAPTER 10
1. Tizard's note recording contents of letter to A.V. Hill, May 31, 1940. Tizard Papers, Imperial War Museum, London HTT 706.
2. Letter A. V. Hill to Sir Henry Tizard, Tizard Papers, Imperial War Museum, London HTT 706.
3. Tizard Papers, Imperial War Museum, London HTT 706.
4. Tizard to ACAC, 8 May 1940, "Summary of the file S4471," Tizard Papers, Imperial War Museum, HTT 706.
5. Tizard Papers, Imperial War Museum, London HTT 706.
6. Tizard Papers, Imperial War Museum, London HTT 706.

7. William Bullitt, *For the President* (Boston: Houghton Mifflin, 1972), 398-389.
8. Breckinridge Long, May 22, in *The War Diary of Breckinridge Long*, ed. Fred L. Israel (Lincoln: University of Nebraska Press, 1966), 113.
9. A. V. Hill, "RDF in Canada and the United States; and a Proposal for a General Interchange of Scientific and Technical Information," June 18, 1940, and "Research and Development for War Purposes in Canada," June 18, 1940, Public Record Office, AVIA 22/2286.
10. Lash, *Roosevelt and Churchill*, 213.
11. David Zimmerman, *Top Secret Exchange* (Montreal: McGill-Queens University Press, 1996), 76.
12. Franklin D. Roosevelt, Radio Address to the Democratic National Convention Accepting the Nomination, July 19, 1940. http://www.presidency.ucsb.edu/ws/index.php?pid=15980.
13. *Royal Society: Papers of PMS Blackett*—Tizard, H.T.; Correspondence of various dates CSAC 63.1.79/J.105 1940–1958.
14. Quoted in Clark, *Tizard*, 256.
15. Tizard Papers, Imperial War Museum, London HTT 706.
16. Churchill, *Their Finest Hour*, 356.
17. Lash, *Roosevelt and Churchill*, 207.
18. Roosevelt's record of cabinet meeting, 2 August 1940, in PL, IV pp 1050-51.
19. Prime Ministerial Minute to Lord Halifax, 9 August 1940, Public Record Office. PREM 3/475/1, folio 15.
20. General Account of Army Radar, Prof. J. D. Cockcroft Hill. P9, Tizard Papers, Imperial War Museum, London HTT 701.
21. Bowen, *Radar Days*, 152.
22. Records of the British Technical and Scientific Mission to the United States, Public Record Office, AVIA 10/1.
23. Churchill, *Their Finest Hour*, 356.

CHAPTER 11

1. Research and Development for War Purposes in Canada, June 17, 1940, Public Record Office, AVIA 22/2286, quoted in Zimmerman, *Top Secret Exchange*, 157.
2. A. V. Hill Report, June 18, 1940, National Archives (NARA), New England Branch, Waltham, Radiation Laboratory records, box 49.
3. http://www.winstonchurchill.org/learn/speeches/speeches-of-winston-churchill/113-the-few.
4. Zimmerman, *Top Secret Exchange*, 99.

5. Quoted in Jennet Conant, *Tuxedo Park* (New York: Simon and Schuster, 2002), 181.
6. Conant, *Tuxedo Park*, 163.
7. Conant, *Tuxedo Park*, 172.
8. Conant, *Tuxedo Park*, 176.
9. Order Establishing the National Defense Research Committee, http://docs.fdrlibrary.marist.edu/psf/box2/a13v01.html.
10. Quoted in Conant, *Tuxedo Park*, 171.

CHAPTER 12
1. Bowen, *Radar Days*, 153.
2. Bowen, *Radar Days*, 154.
3. Bowen, *Radar Days*, 154.
4. Bowen, *Radar Days*, 155.

CHAPTER 13
1. Tizard diary, Sept. 9, 1940, Tizard Papers, Imperial War Museum, London HTT 16.
2. Bowen, *Radar Days*, 152.
3. Bowen, *Radar Days*, 158.
4. Using the share of GDP calculator at http://www.measuringworth.com/ukcompare/.
5. General Account of Army Radar, by Prof. J. D. Cockcroft p. 10, Tizard Papers, Imperial War Museum, London HTT 701.
6. Bowen, *Radar Days*, 159.
7. Bowen, *Radar Days*, 159.
8. General Account of Army Radar, by Prof. J. D. Cockcroft p. 10, Tizard Papers, Imperial War Museum, London HTT 701.
9. General Account of Army Radar, by Prof. J. D. Cockcroft p. 10, Tizard Papers, Imperial War Museum, London HTT 701.
10. Bowen, *Radar Days*, 160.
11. Bowen, *Radar Days*, 163.
12. Bowen, *Radar Days*, 164.
13. Conant, *Tuxedo Park*, 193.
14. Buderi, *Invention that Changed the World*, 88.
15. Bowen, *Radar Days*, 168.
16. Bowen, *Radar Days*, 171.
17. General Account of Army Radar, by Prof. J. D. Cockcroft, Tizard Papers, Imperial War Museum, London HTT 701. GEE and LORAN were both radio navigation systems that later played a significant part in the air war – though neither of them were strictly speaking radar devices.

18. Ed Bowles, quoted in Conant, *Tuxedo Park*, 199.
19. Bowen, *Radar Days*, 192.
20. Bowen, *Radar Days*, 173.

Chapter 14

1. Zimmerman, *Top Secret Exchange*, 146.
2. Constantine FitzGibbon, *The Winter of the Bombs* (New York: W. W. Norton, 1958), 168.
3. Bowen, *Radar Days*, 175.
4. Churchill, *Their Finest Hour*, 489.
5. T. S. Harvey, Diary, November 9, 1940 (British Library Add. Mss 56397), quoted in David Reynolds, *The Creation of the Anglo-American Alliance 1937–41* (London: Europa, 1981), 149.
6. Conant, *Tuxedo Park*, 207.
7. Philip Seib, *Broadcasts from the Blitz* (Washington, D.C.: Potomac, 2006), 80.
8. Franklin Roosevelt, Press Conference, December 17, 1940, http://docs.fdrlibrary.marist.edu/odllpc2.html.
9. Reynolds, *The Creation of the Anglo-American Alliance*, 151.
10. http://en.wikisource.org/wiki/Page:United_States_Statutes_at_Large_Volume_27.djvu/348.
11. Lash, *Roosevelt and Churchill*, 263.
12. Franklin Roosevelt, Press Conference, December 17, 1940, http://docs.fdrlibrary.marist.edu/odllpc2.html.
13. Lash, *Roosevelt and Churchill*, 263.
14. FitzGibbon, *The Winter of the Bombs*, 199.
15. FitzGibbon, *The Winter of the Bombs*, 207.
16. Roosevelt, Fireside Chat, December 29, 1940, http://docs.fdrlibrary.marist.edu/122940.html.
17. Roosevelt, Fireside Chat, December 29, 1940, http://docs.fdrlibrary.marist.edu/122940.html.
18. Roosevelt, Fireside Chat, December 29, 1940, http://docs.fdrlibrary.marist.edu/122940.html.

Chapter 15

1. Bowen, *Radar Days*, 182.
2. Bowen, *Radar Days*, 183.
3. Bowen, *Radar Days*, 129.
4. Buderi, *The Invention that Changed the World*, 113.
5. Conant, *Tuxedo Park*, 69.
6. Bowen, *Radar Days*, 185.
7. Bowen, *Radar Days*, 187.

8. Bowen, *Radar Days*, 188.

Chapter 16

1. Buderi, *The Invention that Changed the World*, 113. James Phinney Baxter, *Scientists Against Time* (Cambridge, Mass.: Little, Brown, 1946), 124.
2. Buderi, *The Invention that Changed the World*, 118.
3. *Radar: A Report on Science at War* (Washington, D.C.: U.S. Government Printing Office, 1945), 23.
4. Denis Robinson interview, quoted in Buderi, *The Invention that Changed the World*, 124. Denis Robinson, an oral history conducted in 1991 by John Bryant, IEEE History Center, New Brunswick, N.J. © 1991 IEEE.
5. Denis Robinson interview, quoted in Buderi, *The Invention that Changed the World*, 124. Denis Robinson, an oral history conducted in 1991 by John Bryant, IEEE History Center, New Brunswick, N.J. © 1991 IEEE.
6. Denis Robinson, an oral history conducted in 1991 by John Bryant, IEEE History Center, New Brunswick, N.J. © 1991 IEEE.
7. Denis Robinson, an oral history conducted in 1991 by John Bryant, IEEE History Center, New Brunswick, N.J. © 1991 IEEE.
8. Rowe, *One Story of Radar*, 115.
9. Rowe, *One Story of Radar*, 116.
10. Bernard Lovell, "The Cavity Magnetron in World War II: Was the Secrecy Justified?" *Notes and Records of the Royal Society of London* 58, no. 3 (Sept. 2004): 283–294, quoted on 286.
11. Lovell, "The Cavity Magnetron in World War II: Was the Secrecy Justified?" 286.
12. http://en.wikisource.org/wiki/Roosevelt%27s_Fireside_Chat,_11_September_1941.
13. Charles A. Beard, *President Roosevelt and the Coming of the War, 1941* (New Haven: Yale University Press, 1948), 140.
14. Franklin D. Roosevelt, "Message to Congress Urging the Arming of American Flag Ships Engaged in Foreign Commerce," October 9, 1941, Department of State *Bulletin*, October 11, 1941. Available at http://www.ibiblio.org/pha/policy/1941/411009a.html.
15. Reynolds, From Munich to Pearl Harbor, 164.
16. Obituary, *New York Times*, February 25, 2010.
17. Frances Perkins, in "Notable New Yorkers," Columbia University Libraries Oral History Research Office. http://www.columbia.edu/cu/lweb/digital/collections/nny/perkinsf/index.html.
18. http://en.wikisource.org/wiki/Roosevelt%27s_Fireside_Chat,_9_December_1941.

Chapter 17

1. *Radar: A Report on Science at War*, 9.
2. Denis Robinson, an oral history conducted in 1991 by John Bryant, IEEE History Center, New Brunswick, N.J. © 1991 IEEE.
3. Bowen, *Radar Days*, 114.
4. http://www.uboat.net/fates/losses/chart.htm; and *Radar: A Report on Science at War*, 24.
5. Rowe, *One Story of Radar*, 146.
6. Rowe, *One Story of Radar*, 146.
7. Winston Churchill, *The Second World War, vol. IV: The Hinge of Fate* (London: Cassell, 1951), 250; and http://www.webofstories.com/play/17811.
8. Lovell, "The Cavity Magnetron in World War II: Was the Secrecy Justified?"
9. Rowe, *One Story of Radar*, 148.
10. Lovell, "The Cavity Magnetron in World War II: Was the Secrecy Justified?"
11. Lovell, "The Cavity Magnetron in World War II: Was the Secrecy Justified?"
12. Quoted in Bernard Lovell, *Echoes of War* (Bristol: Adam Hilger, 1991), 146.
13. Buderi, *The Invention that Changed the World*, 189.
14. *Radar: A Report on Science at War*, 30.
15. Baxter, *Scientists Against Time*, 115.
16. Baxter, *Scientists Against Time*, 87.

Chapter 18

1. Zimmerman, *Top Secret Exchange*, 150.
2. Leonard Bertin, *Atom Harvest* (San Francisco: W. H. Freeman, 1957), 57.
3. Bertin, *Atomic Harvest*, 57.
4. Quoted in Richard Rhodes, *The Making of the Atomic Bomb* (New York: Simon and Schuster, 1986), 372.
5. Quoted in Rhodes, *The Making of the Atomic Bomb*, 373.
6. Bertin, *Atom Harvest*, 67.
7. Records of the British Technical and Scientific Mission to the United States, Public Record Office, AVIA 10/1.
8. http://raf-lincolnshire.info/bombercmd/bombercmd.htm.
9. Baxter, *Scientists Against Time*, 142.
10. Extract from a letter to Henry Tizard from Miss Geary, dated December 28, 1954, Tizard Papers, Imperial War Museum, London HTT 706.
11. Arthur O. Bauer—Centre for German Communication and related technology, *Naxos, The History of a German Mobile Radar Direction Finder, 1943—*

1945. http://www.cdvandt.org/Naxos95nw.pdf.
12. Lovell, "The Cavity Magnetron in World War II: Was the Secrecy Justified?"
13. Lovell, "The Cavity Magnetron in World War II: Was the Secrecy Justified?"
14. Buderi, *The Invention that Changed the World*, 49.
15. Rowe, *One Story of Radar*, 151.
16. Sir Henry Tizard, "Science and the Services," *Journal of the Royal United Services Institute* 91 (1946): 333–344.

Bibliography

Archives

Cherwell Archives, Nuffield College Oxford

National Archives, Public Record Office, Surrey

Tizard Papers, Imperial War Museum, London

Books

Baxter, James Phinney, *Scientists Against Time* (Boston: Little, Brown, 1946).

Beard, Charles A., *President Roosevelt and the Coming of the War, 1941* (New Haven: Yale University Press, 1948).

Berlin, Isaiah, *Mr. Churchill in 1940* (Boston: Houghton Mifflin, 1960).

Bertin, Leonard, *Atom Harvest* (San Francisco: W. H. Freeman, 1957).

Boog, Horst, ed. *The Conduct of the Air War in the Second World War* (New York: Berg, 1992).

Bowen, E. G., *Radar Days* (Bristol: Adam Hilger, 1987).

Buderi, Robert, *The Invention that Changed the World* (New York: Simon and Schuster, 1996).

Bullitt, William, *For the President* (Boston: Houghton Mifflin, 1972).

Bullock, Alan, *Hitler: A Study in Tyranny* (London: Odhams, 1952).

Casey, Steven, *Cautious Crusade* (New York: Oxford University Press, 2001).

Churchill, Winston, *The Second World War, Vol. I, The Gathering Storm* (London: Cassell, 1948).

Churchill, Winston, *The Second World War, Vol. II, Their Finest Hour* (London: Cassell, 1948).

Churchill, Winston, *The Second World War, Vol. IV: The Hinge of Fate* (London: Cassell, 1951).

Clark, Ronald W., *Tizard* (London: Methuen, 1965).

Collier, Basil, *The Battle of Britain* (London: Batsford, 1962).

Conant, Jennet, *Tuxedo Park* (New York: Simon and Schuster, 2002).

Cornwell, John, *Hitler's Scientists* (London: Viking, 2003).

Crowther, J. G., *Statesmen of Science* (London: Cresset Press, 1965).

FitzGibbon, Constantine, *The Winter of the Bombs* (New York: W. W. Norton, 1958).

Friedman, Norman, Naval Radar (Annapolis, MD: Naval Institute Press, 1981).

Israel, Fred L., ed., *The War Diary of Breckinridge Long* (Lincoln: University of Nebraska Press, 1966).

Langer, William, and S. Everett Gleason, *The Challenge to Isolation* (New York: Harper, 1952).

Lash, Joseph P., *Roosevelt and Churchill 1939–1941* (London: Andre Deutsch, 1977).

Lovell, Bernard, "The Cavity Magnetron in World War II: Was the Secrecy Justified?" *Notes and Records of the Royal Society of London* 58, no. 3 (Sept. 2004): 283–294.

Lovell, Bernard, *Echoes of War* (Bristol: Adam Hilger, 1991).

Mowat, C. L., ed., *The New Cambridge Modern History, Vol. XII: The Shifting Balance of World Forces 1898–1945* (Cambridge: Cambridge University Press, 1968).

Nixon, Edgar B., ed., *Franklin D. Roosevelt and Foreign Affairs*, 1st series, 3 vols. (Cambridge Mass.: Belknap Press, 1969).

Radar: A Report on Science at War (Washington, D.C.: U.S. Government Printing Office, 1945)

Reynolds, David, *From Munich to Pearl Harbor* (Chicago: Ivan R. Dee, 2001).

Reynolds, David, *The Creation of the Anglo-American Alliance 1937–41* (London: Europa, 1981).

Rhodes, Richard, *The Making of the Atomic Bomb* (New York: Simon and Schuster, 1986).

Roosevelt, Franklin D., *FDR: His Personal Letters*, ed. Elliott Roosevelt, 2 vols. (New York: Duell, Sloan, and Pearce, 1950).

Roosevelt, Franklin D., *The Public Papers and Addresses of Franklin D Roosevelt*, 2nd series, ed. Samuel I. Rosenman, 13 vols. (New York: Russell and Russell, 1941-50).

Rowe, A. P., *One Story of Radar* (Cambridge: Cambridge University Press, 1948).

Seib, Philip, *Broadcasts from the Blitz* (Washington, D.C.: Potomac, 2006).

Smith, Geoffrey S., "Isolationism, the Devil, and the Advent of the Second World War: Variations on a Theme," *International History Review* 4, no. 1 (Feb. 1982).

Wiltz, John E., *In Search of Peace: The Senate Munitions Inquiry, 1934–1936* (Baton Rouge: Louisiana State University Press, 1963).

Zimmerman, David, *Top Secret Exchange* (Montreal: McGill-Queens University Press, 1996).

Index

Acknowledgments

The list of people who have supported me in the writing of this book is too long to spell out in its entirety here. Some of them will not even know they did so.

My publisher, Bruce H. Franklin, deserves thanks, for having the foresight and bravery to encourage me to write the book in the first place. Lucinda Bartley and Noreen Abel-O'Connor did a terrific job whipping it into shape. Thank you.

I would particularly like to thank Dr. Phil Judkins of the Defence Electronics History Society. He is an enormously rich source of information, and he gave of it generously. The two Davids (Robertson and Phelps) picked me up on linguistic as well as historical errors. As always, the breadth of my brother David's knowledge was invaluable. Peter Hoare and Angela Hind of Pier Productions were the first to see it as a story worth investing in. Bob Spence of Imperial College gave me the chance to flesh it out for the inaugural Peter Lindsay Memorial Lecture. And thank you, Mick Pitt, for putting me on to the story in the first place.

This book was a long time in the making. There are many people whose patience was sorely tested. Principal among them was Tamsen Courtenay, my partner who was always there to encourage me when it seemed overwhelming. My children, Leo and Alice, believed in me (as children should), and gave me the courage to know that I had a story worth telling. Most of all I would like to thank my father, who lived through these events and spent many years trying to interest me in the deeper, richer stories of the Second World War. I hope some of his humanity has infected this book, and I hope, too, that he would have been proud that it makes just one of those stories available to a wider audience.